Lecture Notes in Mathematics

Edited by A. Dold and B. Eckmann

424

Lionel Weiss
Jacob Wolfowitz

Maximum Probability Estimators and Related Topics

Springer-Verlag
Berlin · Heidelberg · New York 1974

Prof. Dr. Lionel Weiss
Department of Operations Research, Upson Hall
Cornell University
Ithaca, NY 14853/USA

Prof. Dr. Jacob Wolfowitz
Department of Mathematics
University of Illinois at Urbana
Champaign
Urbana, Il 61801/USA

Library of Congress Cataloging in Publication Data

Weiss, Lionel, 1923-
Maximum probability estimators and related topics.

(Lecture notes in mathematics ; 424)
Bibliography: p.
Includes index.
1. Probabilities. I. Wolfowitz, Jacob, 1910- joint author. II. Title. III. Series: Lecture notes in mathematics (Berlin) ; 424.
QA3.L28 no. 424 [QA273] 510'.8s [519.2'87] 74-23843

AMS Subject Classifications (1970): 62F05, 62F10, 62F20

ISBN 3-540-06970-4 Springer-Verlag Berlin · Heidelberg · New York
ISBN 0-387-06970-4 Springer-Verlag New York · Heidelberg · Berlin

Offsetdruck: Julius Beltz, Hemsbach/Bergstr.

PREFACE

In the last few years, the authors have, in a number of papers, been developing the theory of maximum probability estimators. This theory is a comprehensive one of asymptotically efficient estimators, and, as the reader will easily verify below, it is not, as has been sometimes incorrectly stated, limited either to the non-regular case or to errors of estimation of order $n^{-1/2}$. On the contrary, it includes both maximum likelihood theory and the non-regular case as special cases. The present monograph is intended as an introduction to the theory which could be studied by graduate students working by themselves or in a seminar. It is largely self-contained, gives some important proofs in greater detail, brings together material from a number of papers, and provides supplementary discussion not present in the papers. The appendix contains a number of non-trivial illustrative examples; more will be found in the original papers. The monograph is devoted almost entirely to the case where θ, the parameter being estimated, is a scalar, but the appendix contains a number of examples where θ has several components. These will cause the reader no difficulty, and it is hoped that they will encourage him to go to the original papers for, inter alia, the multidimensional theory. The monograph does not exhaust the contents of the papers.

Chapter 7 discusses the application of the basic theory to the problem of testing hypotheses. It extends earlier joint work of both authors, but was written by L. Weiss. The rest of the monograph can be read with no reference to Chapter 7.

We thank Dr. Klaus Peters of Springer Verlag for the invitation to publish in this series, and Mrs. Jan Post for typing the manuscript so well. During the preparation of the manuscript the first author was partly supported by the National Science Foundation under Grant No. GP-31430X, and the second author was partly supported by the U.S. Air Force under Grant AF-AFOSR-70-1947, monitored by the Office of Scientific Research. This assistance is hereby gratefully acknowledged.

TABLE OF CONTENTS

CHAPTER 1: PURPOSE OF THIS MONOGRAPH

This monograph is devoted to the method and theory of the estimators called by the authors "maximum probability" estimators. This method will yield asymptotically efficient estimators for almost all statistical problems of theoretical or practical interest. A number of such problems, some of them new, illustrate the theory. The present work is intended to be approximately self-contained, because our work on this subject is now spread over a number of papers, the earlier of which do not have the theory in final form.

Why study asymptotic estimation?

One reason is the history of the problem. Since the discovery, about fifty years ago, of the maximum likelihood (m.ℓ.) estimator by R.A. Fisher, several hundred papers have been written on the subject. The discovery of the m.ℓ. estimator is a truly brilliant one. The m.ℓ. estimator is, without doubt, the most frequently used estimator in statistical practice. Most statisticians, when forced to make a decision in a new problem for which no theory exists, instinctively use an m.ℓ. estimator. Its theory is elegant and attractive. However, even after all these years of work, it still has certain inadequacies (see Section 2 below). A satisfactory completion of m.ℓ. theory is desirable from the mathematical and aesthetic points of view. The theory of maximum probability (m.p.) estimators includes that of m.ℓ. estimators as a very special case. This fact alone is an objective justification for the study of m.p. estimators.

It also is not without "practical" value. Can asymptotic procedures have any practical value? Roughly speaking, sample sizes in statistical problems can be classified as small, medium, and large. Not much can be expected from the small ones. It is those of medium size which occur frequently and for which we would want to use the best possible estimator. The latter is bound to depend in a crucial manner on the distributions and loss function involved. Indeed, most often there will not be an estimator optimal for all values of the parameter (for all states of Nature). The statistician will seek to obtain the class of all admissible estimators, and will try to make some choice (i.e., some compromise) in this class. An asymptotically efficient estimator is simpler, because in the limit the role of the distributions is much less crucial. Thus an asymptotically efficient estimator may be considered as a first approximation to a good estimator for the medium size sample. Perhaps the ultimate solution for medium size samples will begin with an asymptotically efficient estimator and add correcting terms.

Of course, the asymptotic theory is ideal for large samples. Then, when it is known that the actual distribution is to within the desired degree of approximation to the limiting distribution, the m.p. estimator is actually the solution to the problem to within the desired degree of approximation. There are some problems in physics and engineering where it is easy to obtain a large number of observations.

In the next section we will discuss inadequacies of the classical m.ℓ. estimator. Here we want to say that ours is a unified theory of asymptotically efficient estimation, and not just a treatment of the "non-regular" case (see Section 2).

An incidental benefit from our treatment is an intuitive explanation of why the m.ℓ. estimator, indeed, the m.p. estimator itself, is asymptotically efficient. We have long sought for such an explanation (for the m.ℓ. estimator) and one of us tried to give one in [19]. After all, there is no obvious a priori reason why maximizing the likelihood function should yield an asymptotically efficient estimator. To the best of our knowledge no other mathematicians have tried to give such an explanation, but several philosophers have felt the need for one. Several of the latter have made the m.ℓ. estimator the axiomatic basis of their system of estimation, a procedure which is a model method of begging the question.

From several proofs of the consistency of the m.ℓ. estimator (e.g., Wald's [9]) the reasons for the consistency become clear. They are 1) the law of large numbers and 2) Jensen's inequality applied to $E_\theta \left[\frac{\partial \log f(X|\theta)}{\partial \theta}\right]_{\theta=\theta'}$. The reasons for efficiency are another matter.

Of course, we fully realize that many mathematicians feel no need for an intuitive explanation, and that that which one person considers intuitive need not be intuitive to another.

The central problem of statistical inference and the goal of all statistical theory is the development of estimators efficient in some appropriate sense. In the asymptotic theory, efficient estimators are of course consistent. In the present state of statistical knowledge consistent estimators are usually not difficult to give. In most cases they are actually easy to find, and in almost all the other cases (e.g., estimating the parameters of a linear regression with both variables subject to error) they can be found by the minimum distance method (see, for example, [20]). It will appear later that in many non-regular problems the m.p. estimator differs from the m.ℓ. estimator by a constant which is of the same order as the error of estimation. Hence the m.ℓ. estimator cannot be asymptotically efficient, since the m.p. estimator is efficient. Consequently there seems little point, for such problems, in the expenditure of much effort in finding the m.ℓ. estimator and its limiting distribution, unless these are intended to shed light on the m.p. estimator. In the present state of statistical knowledge what is needed for an asymptotic theory is efficient estimators.

Henceforth, throughout this monograph, by "efficient" we will always mean asymptotically efficient.

CHAPTER 2: THE MAXIMUM LIKELIHOOD ESTIMATOR

Let $f(x \mid \theta)$ be the density, at the point x of the real line, of a chance variable X. The density depends upon a parameter θ, unknown to the statistician except for the fact that it belongs to a known set Θ. Strictly speaking, we should have said "a" density, not "the" density, but we abjure such pedantic niceties. Let $X_1, \ldots, X_n$ be independent chance variables with the same density as X. These are the "observed" chance variables. The convention that the X_i's are independent and have the same density as X is to obtain throughout Section 2. The m.ℓ. estimator $\hat{\theta}_n$ is that function of $X_1, \ldots, X_n$ which maximizes, with respect to θ, the likelihood function (of θ) $L_n(\theta) = \prod_{i=1}^{n} f(X_i \mid \theta)$.

Such a maximizing value (in Θ) need not always exist, as an example in [5] shows. If it does exist it need not be unique. It was pointed out by Wald that the argument in [3] proved that a root of the likelihood equation was consistent, without indicating which root was the consistent estimator. Both of these difficulties disappear in almost all cases by using the device introduced in [13] (see also [15]) and discussed in detail in Section 3 below. According to this idea, one maximizes $L_n(\theta)$ with respect to θ in in a suitable neighborhood of any consistent estimator of θ. As we have said earlier, in most problems a consistent estimator can be found easily; the crucial problem is efficiency. Henceforth we assume, unless the contrary is explicitly stated, that $\hat{\theta}_n$ exists and is unique.

Under certain conditions on $f(\cdot \mid \cdot)$, about which more in a moment, one can prove that $\sqrt{n}\,(\hat{\theta}_n - \theta)$ is asymptotically normally distributed, with mean zero and variance $\sigma_\theta^2(\hat{\theta})$, say; we write this by saying that $\sqrt{n}\,(\hat{\theta}_n - \theta)$ is asymptotically $N(0, \sigma_\theta^2(\hat{\theta}))$. The brilliant idea, or, as some may prefer to call it, conjecture, of Fisher was this: Let T_n be any other estimator which is asymptotically $N(0, \sigma_\theta^2(T))$. Then, for every $\theta \in \Theta$,

(2.1) $\sigma_\theta^2(\hat{\theta}) \leq \sigma_\theta^2(T)$.

It is clear that, for the proof of the asymptotic normality of $\sqrt{n}\,(\hat{\theta}_n - \theta)$ and of (2.1), regularity conditions on $f(\cdot|\cdot)$ will be needed. The mere statement of (2.1) requires that $\sqrt{n}\,(T_n - \theta)$ be asymptotically normal. Thus there are regularity conditions on $f(\cdot|\cdot)$ and on $\{T_n\}$. We shall discuss the latter first.

The following example, due to Hodges, shows that (2.1) cannot hold for every T_n and every θ: Let

(2.2) $$f(x \mid \theta) = (2\pi)^{-1/2} \exp\{-(x - \theta)^2/2\}$$

$$\overline{X}(n) = n^{-1} \sum_{i=1}^{n} X_i$$

$$T_n = \overline{X}(n) \quad \text{when} \quad |\overline{X}(n)| > n^{-1/4}$$

$$T_n = 0 \quad \text{when} \quad |\overline{X}(n)| \leq n^{-1/4} .$$

Then $\hat{\theta}_n = \overline{X}(n)$, $\sigma_\theta^2(\hat{\theta}) \equiv 1$, and $\sigma_0^2(T) = 0$. Thus (2.1) is not true. Even more, $\sigma_\theta^2(T) = 1$ for $\theta \neq 0$, so that, for every θ

$$\sigma_\theta^2(\hat{\theta}) \geq \sigma_\theta^2(T)$$

and the equality sign does not hold at $\theta = 0$.

Now a reasonable statistician, dealing with a practical problem, can scarcely be expected to use an estimator like the previous T_n. Thus we can see this example for what it is, a mathematical tour de force. Mathematical statistics is part of applied mathematics, which is a branch of the art of the possible. We all have an intuitive idea of what efficiency is. The problem is to give a definition which expresses this idea well, and then to give a method of obtaining estimators which are efficient in this sense. Thus it is not unreasonable that we should have to put restrictions on the competing estimators T_n so as to exclude artificial estimators, like the one above, which no practical statistician would ever use. The problem is, however, to exclude only artificial competitors. If we exclude sensible and practical competitors then any claims about the optimality of the m.ℓ. or any other estimator are hollow indeed, and the theorems proved do not describe the physical reality and are not of practical value or aesthetic interest.

The requirement that any estimator admitted to competition must be asymptotically normally distributed is an artificial requirement. It is made only for the convenience of the theory, so that we can compare the estimators by their variances. It corresponds to no physical requirement. There is no a priori reason why an efficient estimator should be asymptotically normally distributed. Yet this requirement of asymptotic normality has stood largely unchallenged for the fifty years of the life of the m.ℓ. theory.

Sometimes more is implicitly required. An argument due to Fisher and uncritically repeated by many writers goes like this: Let T_n be an (asymptotically normal) competing estimator and suppose that it has already been proved that $\hat{\theta}_n$ is efficient. Suppose also that

$$\frac{\sigma_\theta^2(\hat{\theta})}{\sigma_\theta^2(T)} = c < 1 . \qquad (c \text{ a function of } \theta)$$

It is then claimed that the correlation coefficient ρ of $\sqrt{n}\,(\hat{\theta}_n - \theta)$ and $\sqrt{n}\,(T_n - \theta)$ in their limiting distribution is $\sqrt{c}$. The "proof" consists in forming the estimator $(1-\Delta)\,\hat{\theta}_n + \Delta T_n$, whose limiting distribution has variance

$$\sigma^2(\hat{\theta})[1 + 2\Delta\,(\frac{\rho}{\sqrt{c}} - 1) + O(\Delta^2)] .$$

If $\rho \neq \sqrt{c}$ then, for Δ small in absolute value and of sign opposite to that of $(\frac{\rho}{\sqrt{c}} - 1)$, the estimator $(1 - \Delta)\,\hat{\theta}_n + \Delta\, T_n$ would be more efficient than $\hat{\theta}_n$. This seeming contradiction is supposed to prove that $\rho = \sqrt{c}$. The fallacy lies in the implicit assumption that $(1 - \Delta)\,\hat{\theta}_n + \Delta\, T_n$ is an admissible competitor to $\hat{\theta}_n$, i.e., that it is asymptotically normally distributed. Even if $\hat{\theta}_n$ and T_n are each asymptotically normal, it does not follow that every linear combination of them is asymptotically normal, unless $\hat{\theta}_n$ and T_n are <u>jointly</u> asymptotically normal.

Paradoxically, it often happens that, when the m.ℓ. estimator is uncritically used in cases which do not fall into the domain of the existing theory (into the so-called "regular" case, see below), the distribution of the m.ℓ. estimator is not asymptotically normal! As an example take the case where

$$(2.3)\qquad f(x \mid \theta) = e^{-(x-\theta)} \quad , \quad x \geq \theta$$
$$= 0 \quad , \quad x < \theta \,.$$

The m.ℓ. estimator is $\min (X_1, \ldots, X_n)$ and is not, after proper normalization, asymptotically normal.

We emphasize that the statistical problem is always that of finding efficient estimators. There are many books and papers which obtain the asymptotic distribution of the m.ℓ. estimator in cases where the m.ℓ. estimator has not yet been proved efficient. Unless the latter is ultimately done, the value of these results may be marginal. The crucial question is always that of finding efficient estimators. In many, though not all, problems it is relatively easy to give a consistent estimator.

Let us now discuss the regularity conditions on $f(\cdot|\cdot)$ needed for (2.1). The inequality (2.1) has been studied under different regularity conditions, all of which, however, have much in common. In the literature those of [3] are often cited as an example. The problem treated is usually referred to as the "regular" case. The recent book [8] is a very readable and comprehensive treatment of the regular case. A feature of the regular case is that the measures which correspond to different θ are all absolutely continuous with respect to each other. It is usually difficult to make the different sets of conditions directly comparable. Of course, while it is desirable to have the regularity conditions as weak as possible, we are not now interested in slight improvements of the regularity conditions, because our present concern is with a major defect in all such sets of regularity conditions. This defect is that they arbitrarily exclude even from consideration many important and interesting problems and distributions. For example, the density (2.3) is not "regular." Neither is the density

$$(2.4) \qquad f(x \mid \theta) = 1/2 , \quad |x - \theta| \leqq 1$$
$$= 0 , \quad |x - \theta| > 1 .$$

Neither are many other such densities, about which there is nothing irregular. Neither is any density for which it is possible to estimate θ more closely than to within an error of $O_p(\frac{1}{\sqrt{n}})$. Thus the term "regular" for the conditions under which the m.ℓ. estimator is efficient is more of a mathematical trick than anything which truly corresponds to the ordinary connotation of the word regular.

Whatever results have been proved about the optimality of the m.ℓ. estimator apply only to the cases where the X_i's are independent and identically distributed, or where the dependence among the X_i's is relatively simple (e.g., they form a Markov chain). It would be desirable to have a theory which can treat more complex cases of dependence.

It would also be desirable to be able to treat more adequately the case where the dimension m of θ is > 1. So far we have written only about the case $m = 1$. Obviously this is only one case of importance.

To summarize, the inadequacies of m.ℓ. theory are: 1) unreasonable restrictions on the competing estimators which diminish the value of results about the optimality of the m.ℓ. estimator 2) restriction to the "regular" case, which has little if anything to do with regularity except the pretentious name, and which arbitrarily excludes many, many problems of importance 3) inadequate treatment of cases of dependent observations 4) inadequate treatment of cases where $m > 1$.

Before proceeding to the m.p. estimator we digress for a word about the inequality known as the Cramér-Rao, although it was first discovered by Fréchet. As is well known, this inequality says that, under certain conditions, any estimator T_n satisfies

$$(2.5) \qquad E_\theta(T_n - E_\theta T_n)^2 \geq \frac{\left[1 + \frac{db(\theta)}{d\theta}\right]^2}{n\, E_\theta \left[\frac{\partial \log f(X|\theta)}{\partial\theta}\right]^2}$$

where $b(\theta)$ is the bias, at θ, of T_n. This inequality is not really relevant to an asymptotic theory, because it deals, not with the limiting distribution, but the finite sample size distribution. It is obtained in the regular case with additional restrictions on the estimators considered. If the distribution of the estimator is very different from normal the variance may be a poor estimator of the scatter. The term $\frac{db}{d\theta}$ in the numerator diminishes considerably the value of the inequality, and there is no compelling reason to limit one's self to unbiased estimators. Indeed, the m.ℓ. estimator is often biased. Thus the Fréchet inequality is irrelevant to our present purposes.

CHAPTER 3: THE MAXIMUM PROBABILITY ESTIMATOR

The outline of this section is as follows: 1) definition of the m.p. estimator 2) statement of the theorem which asserts its efficiency 3) discussion of the assumptions 4) proof of the theorem 5) a few remarks about the intuitive significance of the proof and some properties of the estimator 6) relation to the m.ℓ. estimator 7) extension to other cases 8) a few examples.

3(1) For every large n let $X(n)$ denote the vector of observed chance variables. The estimator is to be a Borel measurable function of $X(n)$. In Section 2 $X(n)$ was $(X_1,\ldots,X_n)$. Now, and in general, $X(n)$ need not have n components, nor need its components be independently and identically distributed. Let $K_n(x|\theta)$ be the density, with respect to a σ-finite measure μ_n, of $X(n)$ at the point x of the space of $X(n)$, when θ is the, unknown to the statistician, value of the parameter. It is known to the statistician that θ is in the known set Θ, which is called the parameter space. Θ is assumed to be an open subset of Euclidean m-space. We assume that $K_n(x|\theta)$ is jointly measurable in (x,θ) with respect to μ_n x Borel measure, and that, for either x or θ fixed, $K_n(x|\theta)$ is a measurable function in the other variable. In this monograph, except where the contrary is explicitly stated, we proceed as if $m = 1$. This restriction is not necessary and is made only for the sake of simplicity. The reader is referred to our references for details and examples of cases where $m > 1$; the details of the theory for $m > 1$ will usually be obvious. Some examples with $m > 1$ will be given in this monograph.

Let $k(n) > 0$, $k(n) \uparrow \infty$, be a normalizing factor for the family $K_n(\cdot|\theta)$, $n = 1,2,\ldots$; $\theta \in \Theta$. Intuitively speaking, this means that the <u>best</u> that any estimator T_n which is a function of $X(n)$ can estimate θ is to within $0_p(\frac{1}{k(n)})$. Of course, this is not a precise definition and we proceed to give one now, but the intuitive description will be very useful. The function $k(\cdot)$ is a normalizing factor for the family $K_n(\cdot|\theta)$, $n = 1,2,\ldots$; $\theta \in \Theta$, if the following conditions are satisfied for any open subset Θ_0 of Θ:

a) There exists an estimator $T_n^{(1)}$ with the following property: Let $\varepsilon > 0$ be arbitrary. For each $\theta \in \Theta_0$ there exists $M^{(1)}(\theta) > 0$ such that

$$\underline{\lim_{n\to\infty}} P_\theta\{|k(n)(T_n^{(1)} - \theta)| < M^{(1)}(\theta)\} > 1 - \varepsilon .$$

For every compact θ-set the function $M^{(1)}(\cdot)$ is bounded and the approach to the lim inf is uniform.

b) Let $k'(n)$ be a function such that $\lim_{n\to\infty} \frac{k'(n)}{k(n)} = \infty$. Let T_n be any estimator and $M > 0$ any number. There exists a $\theta \in \Theta_0$ such that

$$\lim_{n\to\infty} P_\theta\{|k'(n)(T_n - \theta)| < M\} = 0 .$$

In the regular case (to which m.ℓ. theory applies) $k(n)$ can be $\sqrt{n}$ or $3\sqrt{n}$ or $\sqrt{n} + \log n$. Obviously, two normalizing factors $k(n)$ and $k'(n)$ are essentially equivalent if

$$\lim_{n\to\infty} \frac{k(n)}{k'(n)} = \text{a positive constant}, \tag{3.1}$$

and essentially different if this is not so.

We assume that for the family $K_n(\cdot|\theta)$, $n = 1,2,\ldots$; $\theta \in \Theta$, there is a normalizing factor. This need not always be the case. Thus, for example, the family could have one normalizing factor for odd n and an essentially different one for even n. In this case we are dealing with at least two different problems. As another example, it could be that $\Theta = \Theta_1 \cup \Theta_2$, with $\Theta_1 \cap \Theta_2 = \phi$, Θ_1 and Θ_2 both open, and such that, for θ in Θ_1, there is one normalizing factor, and for θ in Θ_2, an essentially different one. In that case we can consider that we are dealing with two problems, the first being to decide whether the unknown parameter belongs to Θ_1 or Θ_2, the second to estimate θ. It seems to us that the assumption about the existence of a normalizing factor corresponds to actual statistical problems. The methods developed below will also enable us to treat not too pathologic problems where more than one normalizing factor is involved.

Let R be a bounded, Borel measurable subset of the real line. Later we will show how to remove the restriction of boundedness. R is at the disposal of the statistician who chooses it for each problem; it is closely connected with the loss function or utility function for the problem. Its precise meaning will appear from the theorem below. Define the set $\frac{R}{k(n)}$ as follows: a point t is in $\frac{R}{k(n)}$ if and only if $k(n)t$ is in R. Define the set $\{d - \frac{R}{k(n)}\}$ as follows: a point t is in this set if and only if $d - t$ is in $\frac{R}{k(n)}$. The idea is that the statistician considers it good when $k(n)(T_n - \theta)$ (T_n is the estimator being considered) falls into the set R, and bad when it does not. Thus R really defines a $0 - 1$ loss function; more general loss functions will be considered in Section 4 below. The introduction of R, obvious as it is, is completely new in asymptotic estimation theory and leads to an essential advance.

A particularly simple and important R is the interval $(-r,r)$, centered at the origin. In the regular case, with $k(n) = \sqrt{n}$, this means that the statistician accounts it a success when T_n differs from θ by not more than $\frac{r}{\sqrt{n}}$ in absolute value. What may be unreasonable and hence undesirable, is the discontinuity between what happens when $|T_n - \theta|$ is less than, or greater than, $\frac{r}{\sqrt{n}}$. Thus a continuous loss function may be more desirable, and, as we have said, we treat this in Section 4. In some cases what will be important to the statistician will be not the absolute error $|T_n - \theta|$ but the error relative to θ. For example, the statistician may account it a success only when $\frac{-b_1\theta}{k(n)} < T_n - \theta < \frac{b_2}{k(n)}$, with b_1 and b_2 positive. This calls for an R which depends upon θ; we discuss this at the end of this section.

Before proceeding to define the m.p. estimator, or, more properly, an m.p. estimator, we must point out an obvious fact about all asymptotically efficient estimators. They can never be unique, and always what is defined is an equivalence class of them. For example, the estimator $\hat{\theta}_n + \frac{1}{\sqrt{n}\log n}$ possesses the same asymptotic properties as $\hat{\theta}_n$ in the regular case. In our more general problem, any two estimators T_n and T'_n, such that, for every θ in Θ,

$$\text{(3.2)} \qquad \lim_{n\to\infty} [P_\theta\{k(n)(T_n - \theta) \in R\} - P_\theta\{k(n)(T'_n - \theta) \in R\}] = 0 ,$$

are (asymptotically) equivalent. To avoid circumlocutions we shall frequently or always speak of "the" such and such an estimator, when what is strictly meant is "an" estimator.

We now define the m.p. estimator Z_n (with respect to R). Always Z_n depends on R, but most of the time we shall not find it necessary to mention R explicitly and will not do so. Z_n is (a chance variable which is) that value of d for which the integral

$$\text{(3.3)} \qquad \int K_n(X(n)|\theta)d\theta ,$$

over the set $R_n(d) = \{\theta \mid \theta \in d - \frac{R}{k(n)}\}$, is a maximum.

In some problems the maximizing value may not always exist. When this is so we change the above definition slightly as follows: Let $\ell_n \to 0$, $\ell_n > 0$. Let Z_n be a chance variable such that

$$(3.4) \qquad \int_{R_n(Z_n)} K_n(X(n)|\theta)d\theta > (1 - \ell_n)\sup_d \int_{R_n(d)} K_n(X(n)|\theta)d\theta .$$

It will easily be seen below that the optimal properties proved for Z_n defined according to the first definition hold for Z_n defined in this slightly modified manner, provided that always

$$\sup_d \int_{R_n(d)} K_n(X(n)|\theta)d\theta \neq \infty .$$

Even if this is not so, in most cases one can use the method described in the next paragraph (according to which one maximizes the integral with respect to d in a small sphere (actually an interval for $m = 1$) there described).

What if Z_n is not unique? In most cases one can proceed as follows: Let $\phi_n = \phi_n(X(n))$ be a consistent estimator of θ such that $\phi_n - \theta = O_p([k(n)]^{-1})$ (for all $\theta \in \Theta$, of course). Let $k_1(n) \to \infty$ in such a way that $h_1(n) = \frac{k_1(n)}{k(n)} \to 0$. Define Z_n (a chance variable) as a value of d (usually unique) which maximizes the integral (3.3) with respect to d in the closed "sphere" centered at ϕ_n and of radius $h_1(n)$. We will show in Section 3.9 that, when Z_n is so defined, Theorem 3.1 still holds.

[In most problems an estimator with the properties of ϕ_n is easy to find. There are exceptions; e.g., [7], [18], and [20]. Since the radius of the above sphere (i.e., $h_1(n)$) approaches zero it is very reasonable to expect that, for large n, there will be only one maximum in the sphere. Also, $k_1(n)$ may be chosen so that

$k_1(n) \to \infty$ very slowly, although this may have certain disadvantages. The condition that $\phi_n - \theta = O_p([k(n)]^{-1})$ is not essential, but then $k_1(n)$ must fulfill a different condition (see end of Section 3.9). For example, in the regular case with the density (2.2), $M_n + \frac{1}{\log n}$ is consistent, where M_n is the median of $X_1,\ldots,X_n$. Since $M_n - \theta = O_p(\frac{1}{\sqrt{n}})$, it follows that $\bar{X}(n) - M_n - \frac{1}{\log n} = O_p(\frac{1}{\log n})$. If $k_1(n) = n^{1/4}$, $\bar{X}(n)$ need not lie in the interval of half-length $n^{-1/4}$ about $M_n + \frac{1}{\log n}$.]

To simplify matters we henceforth assume, unless the contrary is explicitly stated, that the chance variable Z_n is the unique value of d for which the integral (3.3) over the set $R_n(d)$ is a maximum, except perhaps on a set of μ_n-measure zero. Usually, when this is not so, a simple modification of the theory below will be sufficient for its application.

3(2) From now on we write θ_0 as the "true" (actual) value of θ, and use θ as a running coordinate. The conditions will be stated in terms of θ_0. Since θ_0 is unknown to the statistician the conditions must be met for all possible θ_0 (in Θ). This convention makes for a little more notational simplicity; we borrow this procedure from Wald ([9]). We shall say that a sequence $\{\theta_n, n = 1,2,\ldots\}$ of real numbers is in $H(h)$ if $k(n)|\theta_n - \theta_0| \leq h$ for $n = 1,2,\ldots$. The definition thus depends implicitly on θ_0 and $k(n)$.

<u>Theorem</u> 3.1 Let $k(n)$ be a normalizing factor for the family $K_n(\cdot|\theta)$, $n = 1,2,\ldots; \theta \in \Theta$. Let Z_n be an m.p. estimator with respect to (bounded) R such that:

(3.5) For any $h > 0$ we have, for $\{\theta_n\}$ in $H(h)$,

$$\lim_{n\to\infty} P_{\theta_n}\{k(n)(Z_n - \theta_n) \in R\} = \beta(\theta_0), \text{ say.}$$

(3.6) Let ε and $\delta > 0$ be arbitrary. For h sufficiently large we have, for $\{\theta_n\}$ in $H(h)$,

$$\varliminf_{n\to\infty} P_{\theta_n}\{|k(n)(Z_n - \theta_n)| < \delta h\} \geqq 1 - \varepsilon .$$

Let T_n be any (competing) estimator such that,

(3.7) For any $h > 0$ we have, for $\{\theta_n\}$ in $H(h)$,

$$\lim_{n\to\infty} [P_{\theta_n}\{k(n)(T_n - \theta_n) \in R\} - P_{\theta_0}\{k(n)(T_n - \theta_0) \in R\}] = 0 .$$

Then

(3.8) $$\varlimsup_{n\to\infty} P_{\theta_0}\{k(n)(T_n - \theta_0) \in R\} \leqq \beta(\theta_0) .$$

3(3) Before proceeding to the simple proof of the theorem we shall briefly discuss the form and content of its assumptions.

The assumptions given are not on $K_n(\cdot|\cdot)$ but on the behavior of Z_n ((3.5), (3.6)). One could give conditions on $K_n(\cdot|\cdot)$ which would result in (3.5) and (3.6). These conditions would have to be very complicated in order to apply in the desired generality. Even in the regular case (essentially the only one treated in the literature) the conditions given in the literature usually cover an entire printed page (e.g., see Chapter 6 below) and may not be easy to verify. By contrast, it is usually much easier to verify these conditions for Z_n. The theory of Section 5 below will make easier the computation of Z_n and the verification of (3.5) and (3.6). Extrapolating from our experience in verifying (3.5) and (3.6), we conjecture that (3.5) and (3.6) hold for all "practical" problems and all problems of the statistical literature, and reasonable R.

The condition (3.5) is a little weaker than uniform convergence of $P_\theta\{k(n)(Z_n - \theta) \varepsilon R\}$ to a function of θ on a small closed interval which contains θ_0 in its interior. Uniform convergence is entirely in the spirit of asymptotic theory. Asymptotic theory would be used when the sample size is large, and then the statistician would proceed as if the limit distribution were the actual distribution. If the convergence were not uniform then, since θ_0 is unknown to the statistician, the latter could not possibly know whether the sample size was sufficiently large for the actual distribution to be within the desired degree of accuracy of the limiting distribution. (See also [21].) The condition (3.5) is the one actually used in the proof; uniform convergence is the statistically meaningful condition.

Consider again the estimator in (2.2) in the light of the previous remarks. For this estimator T_n and for every n, there is a point θ_n such that $E_{\theta_n}(T_n - \theta_n)^2$ is actually of order $n^{-1/2}$, whereas, for any reasonable estimator T_n', $E_\theta(T_n' - \theta)^2 = O(n^{-1})$ for every θ.

It is difficult to imagine a statistician's actually using an estimator which does not satisfy (3.7). This intuitive feeling has a sound theoretical basis. The existence of consistent estimators depends upon the continuity in θ of $K_n(\cdot|\theta)$. Consider again the example

$$f(x|\theta) = \frac{1}{\sqrt{2\pi}} e^{-(x-\theta)^2/2} .$$

Let $\Theta = (-1,1)$. Suppose that the point 0 is now renamed 2. Then any estimator of θ which is to be consistent for all θ in the new Θ can be achieved only by means of a device like that in

(2.2). If, in addition, we ask that the estimator be efficient, then it should decide efficiently between the null hypothesis that $\theta = \theta_0$ against the alternative $\theta = \theta_0 + \frac{h}{k(n)}$ or $\theta_0 - \frac{h}{k(n)}$. Hodges' example in (2.2) will not do that.

The condition (3.6) seems to us to be in the same spirit. It is actually verifiable in at least most of the problems in the statistical literature for reasonable R.

3(4) We now prove Theorem 3.1. Let z be the half-width of some interval, centered at the origin, which contains R (bounded). Let $h > z$ be large; we will specify the size of h shortly. Consider the following Bayes problem for each n: The parameter θ is a chance variable, uniformly distributed over the interval $A_n = \{\theta \mid |k(n)(\theta - \theta_0)| \leq h\}$. The gain to the statistician is $+1$ when the estimator lies in the set $(\theta + \frac{R}{k(n)})$, and 0 when this is not so. When a maximizing value does not exist we let the Bayes estimator B_n maximize to within ℓ_n, $\ell_n \to 0$, as in (3.4). Then $B_n(X(n))$ is a chance variable such that

(3.9)
$$\frac{k(n)}{2h}\int_{A_n} P_\theta\{k(n)(B_n-\theta)\varepsilon R\}d\theta \geq \frac{(1-\ell_n)k(n)}{2h}\int_{A_n} P_\theta\{k(n)(T_n-\theta)\varepsilon R\}d\theta ,$$

for any estimator T_n. (In the interests of simplicity of exposition we allow ourselves the abuse of notation involved in using θ as the variable of integration.) The left member of (3.9) can be written as

$$(3.10)\quad \frac{k(n)}{2h}\int_{A_n}\int_{\{x|B_n(x)\varepsilon\theta+\frac{R}{k(n)}} K_n(x|\theta)\mu_n(dx)d\theta = \frac{k(n)}{2h}\int\int_{R_n(B_n)\cap A_n} K_n(x|\theta)d\theta\mu_n(dx) .$$

When $|k(n)(a - \theta_0)| \leqq h - z$, $R_n(a) \subset A_n$. Hence, from the definition of Z_n, it follows that, whenever

$$(3.11) \qquad |k(n)(Z_n - \theta_0)| \leqq h - z ,$$

we may set $B_n = Z_n$.

Now suppose that the theorem is not true and that the left member of (3.8) exceeds $\beta(\theta_0)$ by $4\gamma > 0$. Choose h so large that

$$(3.12) \qquad P_{\theta_n}\{|k(n)(Z_n - \theta_n)| \leqq \gamma h - z\} > 1 - \gamma ,$$

for $\{\theta_n\}$ in $H(h)$ and n sufficiently large. That this can be done follows from (3.6). Define the interval

$$A'_n = \{\theta \mid |k(n)(\theta - \theta_0)| \leqq h(1 - \gamma)\} .$$

For every θ in A'_n the event

$$(3.13) \qquad \{|k(n)(Z_n - \theta)| \leqq \gamma h - z\}$$

implies the event (3.11). It follows from (3.9) and (3.12) that, for n large,

$$(3.14) \qquad \frac{k(n)}{2h}\int_{A_n} P_\theta\{k(n)(Z_n-\theta)\varepsilon R\}d\theta+3\gamma \geqq \frac{k(n)}{2h}\int_{A_n} P_\theta\{k(n)(T_n-\theta)\varepsilon R\}d\theta .$$

From (3.5), (3.7), and (3.14) we obtain

$$(3.15) \qquad \beta(\theta_0) + 3\gamma \geqq \overline{\lim}\ P_{\theta_0}\{k(n)(T_n-\theta_0)\varepsilon R\} .$$

This contradicts the definition of γ and proves the theorem.

3(5) The proof just completed is very simple and perspicuous and provides an intuitive explanation of why the m.p. estimator is efficient. (An adequate intuitive explanation for the efficiency of even the m.ℓ. estimator has been lacking, except perhaps for the argument in [19]; we will see below how the m.ℓ. estimator is a special case of the m.p. estimator.) Even with modest assumptions one would expect that the Bayes estimator with respect to an a priori distribution, uniform on a small interval (of length of order $\frac{1}{k(n)}$) centered at the true value θ_0, would be efficient. But how to obtain such a Bayes estimator, when θ_0 is unknown and is actually the value of the parameter to be estimated? Our proof shows that, asymptotically, the m.p. estimator is this Bayes estimator. It is this fact which explains the efficiency of the m.p. estimator.

Suppose S_n is a statistic sufficient for $\theta \varepsilon \Theta$. This means that $K_n(x|\theta)$ can be written as $K_n^{(1)}(S_n(x)|\theta) \cdot K_n^{(2)}(x)$, $K_n^{(1)} \geqq 0$, $K_n^{(2)} \geqq 0$, $K_n^{(1)}(S_n(x)|\theta)$ measurable for fixed θ, $K_n^{(2)}$ integrable with respect to μ_n and equal to zero for almost all (μ_n) x such that $K_n(x|\theta)$ is zero for all $\theta \varepsilon \Theta$. There is a more intuitive definition to which the former is equivalent, but the one given suits our purpose. It follows immediately that Z_n is a function of $S_n(X(n))$, because $K_n^{(2)}$ does not depend on θ and is therefore constant in the integral (3.3).

Suppose that $K_n(x|\theta) = K_n(x + \vec{y}|\theta + y)$ for every real y, where $\vec{y}$ is the vector in the space of $X(n)$ with all components y. This means that we are dealing with a translation parameter. We now verify easily that $Z_n(x + \vec{y}) = y + Z_n(x)$, as follows:

$$\int_{R_n(d)} K_n(x+\vec{y}|\theta)d\theta = \int_{R_n(d)} K_n(x|\theta-y)d\theta = \int_{R_n(d-y)} K_n(x|\theta)d\theta \; .$$

Hence $Z_n(x + \vec{y}) = y + Z_n(x)$.

3(6) Since the m.ℓ. estimator is so well known and so much has been written about it, we first assume that the reader is familiar with it and may prefer a heuristic treatment because of the complexity of the regularity conditions. A rigorous treatment will be found in Chapter 6 below.

Let $X_1,\ldots,X_n$ be independent with the common density $f(\cdot|\theta)$, which satisfies the conditions of the regular case. It is proved in many books that, with probability approaching one as $n \to \infty$, for all θ which are within $\frac{d}{\sqrt{n}}$ of $\hat{\theta}_n$, d a constant,

(3.16)

$$\sum_{i=1}^{n} \log f(X_i|\theta) \sim \sum_{i=1}^{n} \log f(X_i|\hat{\theta}_n) + \sum_{i=1}^{n} \left.\frac{\partial^2 \log f(X_i|\theta)}{\partial\theta^2}\right|_{\hat{\theta}_n} \frac{(\theta-\hat{\theta}_n)^2}{2} .$$

The symbol "$\sim$" can be understood to mean that the next term in the finite Taylor series can be neglected. For precisely what this means see Chapter 6. Since the left member of (3.16) is $\log K_n(X(n)|\theta)$ it follows from the integral (3.3) and the definition of Z_n that $Z_n = \hat{\theta}_n$ when

(3.17) $\quad R = (-r,r)$

is any interval centered at the origin.

If we are in the regular case it follows that $\sqrt{n}\,(Z_n - \theta_0)$ is in the limit normally distributed with mean zero and variance $\sigma^2_{\theta_0}(\hat{\theta})$. Consider now any competing estimator T_n which satisfies (3.7) and which is such that $\sqrt{n}\,(T_n - \theta_0)$ is asymptotically normal with mean zero (as m.ℓ. theory requires) and variance $\sigma^2_{\theta_0}(T_n)$. It follows from (3.8) and (3.17) that

(3.18) $\sigma^2_{\theta_0}(\hat{\theta}) \leqq \sigma^2_{\theta_0}(T)$,

i.e., the classical statement of efficiency of the m.ℓ. estimator.

We note that, in the regular case, $Z_n = \hat{\theta}_n$ is independent of R for R given by (3.17). If $R = (a,b)$ is not centered at the origin, $Z_n = \hat{\theta}_n + \frac{a+b}{2\sqrt{n}}$.

For the sake of simplicity the theorems of this monograph are proved only for $m = 1$, but they are valid for general m. In the regular case, when R is any bounded convex set in m-space, symmetric about the origin, $Z_n = \hat{\theta}_n$. This follows from a consequence of a theorem of Anderson [1] which we now state. Let R be as above. Then the integral of an m-dimensional normal density whose means are all zero, over the set $d + R$, is a maximum with respect to d when $d = (0,\ldots,0)$. Now, from an expansion like that of (3.16) we conclude that K_n behaves, in the neighborhood of $\hat{\theta}_n$, like an m-dimensional normal density with means given by the vector $\hat{\theta}_n$. Hence, applying the above consequence of Anderson's theorem, we obtain that $Z_n = \hat{\theta}_n$. From the m-dimensional analogue of Theorem 3.1 we conclude that $\hat{\theta}_n$ is efficient with respect to any such R. This last result was first proved, by a different method, by Kaufman [4]. Thus Kaufman's theorem is a special case of Theorem 3.1 (for general m).

3(7) We now ask whether Theorem 3.1 can be extended to sets R which depend upon θ? Such a problem could easily arise as follows: Let θ be a scale parameter, and suppose the statistician wants the relative error to be small, e.g., he wants to maximize the limit of the probability that

(3.19) $$-\frac{b_1\theta}{k(n)} < T_n - \theta < \frac{b_2\theta}{k(n)} ,$$

where b_1 and b_2 are positive constants. Thus $R(\theta) = (-b_1\theta, b_2\theta)$.

Write

$$(3.20) \qquad R_n^*(d) = \{\theta \mid \theta \in d - \frac{R(\theta)}{k(n)}\} ,$$

and let us try to make the proof of Theorem 3.1 go through, with $R_n(d)$ replaced by $R_n^*(d)$. The inner integral in the right member of (3.10) is now over the set $A_n \cap R_n^*(B_n(x))$. Suppose $\bigcup_\theta R(\theta) \subset (-z,z)$. Then, when $|k(n)(a - \theta_0)| \leqq h - z$, $R_n^*(a) \subset A_n$. We now define Z_n (with respect to $\{R(\theta), \theta \in \Theta\}$) as in (3.3), except that $R_n(d)$ is replaced by $R_n^*(d)$. Just as in Theorem 3.1, it follows from the definition of Z_n that, whenever (3.11) holds, we may set $B_n = Z_n$. The remainder of the proof goes through, and we have proved

<u>Theorem</u> 3.2 Let $k(n)$ be a normalizing factor for the family $K_n(\cdot|\theta)$, $n = 1,2,\ldots$, $\theta \in \Theta$. Let the m.p. estimator with respect to $\{R(\theta), \theta \in \Theta\}$, be defined as in (3.3), with $R_n(d)$ replaced by $R_n^*(d)$. Suppose that $\bigcup_{\theta\in\Theta} R(\theta)$ is bounded, and that conditions (3.5)-(3.7), with R replaced by $R(\theta)$, are fulfilled by Z_n and the competing estimator T_n. Then (3.8) holds, with R replaced by $R(\theta)$.

The above results could also be extended to the case where R (or $R(\theta)$) depends on n, not just as above by the affine transformation $\frac{1}{k(n)}$, but more generally. We leave this to the reader.

If $\bigcup_{\theta\in\Theta} R(\theta)$ is not bounded, Theorem 3.2 will still hold if, in the union operation, Θ is replaced by the closed sphere centered at θ_0 and of radius r, where r is any positive number, and the new union is bounded. This is easy to see from the proof of Theorem 3.1.

Suppose now that 1) $\Theta = (0,\infty)$, 2) for every $c > 0$, $\frac{R(c\theta)}{c} = R(\theta)$, and that 3) for every x, $K_n(cx|c\theta) = K_n(x|\theta)$. We will show that then $Z_n(cx) = cZ_n(x)$. This is a consequence of the following:

$$\int_{R_n^*(d)} K_n(cx|\theta)d\theta = \int_{R_n^*(d)} K_n(x|\tfrac{\theta}{c})d\theta = c\int K_n(x|\theta)d\theta ,$$

where the last integral is over the set

$$\{\theta|\theta \in \frac{d}{c} - \frac{R(c\theta)}{ck(n)} = R_n^*(\tfrac{d}{c})\} .$$

3(8) Probably the simplest thing to do when R is unbounded is to use a limiting procedure. Write $R_z = R \cap (-z,z)$, and let $Z_n^{(z)}$ be the m.p. estimator with respect to R_z. The following theorem will suffice us, although it is not the most general possible.

<u>Theorem</u> 3.3 Let $0 < z_1 < z_2 < \cdots \to \infty$ be a sequence for which

$$(3.21)\qquad \lim_{i\to\infty}\lim_{n\to\infty} P_{\theta_0}\{k(n)(Z_n^{(z_i)} - \theta_0)\varepsilon R_{z_i}\} = \lim_{n\to\infty} P_{\theta_0}\{k(n)(Z_n-\theta_0)\varepsilon R\} .$$

Suppose that, for R_{z_i}, $i = 1,2,\ldots,$ the conditions of Theorem 3.1 are fulfilled, with a normalizing factor $k(n)$, by $Z_n^{(z_i)}$ and a competing estimator T_n. Suppose also that

$$(3.22)\qquad \overline{\lim_{i\to\infty}}\ \overline{\lim_{n\to\infty}}\ P_{\theta_0}\{k(n)(T_n-\theta_0)\varepsilon R_{z_i}\} = \overline{\lim_{n\to\infty}}\ P_{\theta_0}\{k(n)(T_n-\theta_0)\varepsilon R\} .$$

Then

$$(3.23) \qquad \lim_{n\to\infty} P_{\theta_0}\{k(n)(Z_n-\theta_0)\varepsilon R\} \geqq \overline{\lim_{n\to\infty}}\, P_{\theta_0}\{k(n)(T_n-\theta_0)\varepsilon R\} .$$

From Theorem 3.1 we obtain, for any i,

$$(3.24) \qquad \lim_{n\to\infty} P_{\theta_0}\{k(n)(Z_n^{(z_i)} - \theta_0)\varepsilon R_{z_i}\} \geqq \overline{\lim_{n\to\infty}}\, P_{\theta_0}\{k(n)(T_n-\theta_0)\varepsilon R_{z_i}\} .$$

We now pass to the limit with i in both members of (3.24). The desired result follows from (3.21) and (3.22).

An undesirable feature of the hypothesis of Theorem 3.3 is that (3.21) and (3.22) have to hold for the same sequence $\{z_i\}$. In most or all problems of importance, (3.21) will hold for any ascending sequence $\{z_i\}$. Condition (3.22) is then not so unreasonable.

3(9) In Section 3(1) we mentioned that one could obtain Z_n by maximizing the integral (3.3) (over $R_n(d)$) with respect to d in the closed sphere centered at the estimator $\phi_n(X(n))$ and of radius $h_1(n)$. (It is required that $|\phi_n(X(n)) - \theta_0| = O_p(\frac{1}{k(n)})$.) Let us prove this now. With probability approaching one, θ_0 lies in a sphere of radius $\frac{1}{2}h_1(n)$ about ϕ_n. The rest of our argument is conditioned upon the latter event. Hence the sphere of radius $h_1(n)$ about ϕ_n includes the sphere of radius $\frac{1}{2}h_1(n)$ about θ_0. It follows from (3.9) that

$$(3.25) \qquad k(n)|B_n - \theta_0| \leqq h + z < \tfrac{1}{2}k_1(n)$$

for large n. Hence, from the present definition of Z_n,

(3.26)

$$\int_{R_n(Z_n)} K_n(x|\theta)d\theta \geqq \int_{R_n(B_n(x))} K_n(x|\theta)d\theta \geqq \int_{R_n(B_n(x))\cap A_n} K_n(x|\theta)d\theta .$$

As before, whenever (3.11) holds, we may set $B_n = Z_n$, and the argument of Theorem 3.1 continues as before. We also note that, as mentioned in Section 3.1 for the original definition of Z_n, we may also maximize in the sphere of radius $h_1(n)$ about ϕ_n only to within $1 - \ell_n$, $\ell_n \to 0$.

[The fact that, in maximizing the integral (3.3), we may limit ourselves to values in the small sphere described above, makes it more plausible that condition (3.6) is satisfied in most problems of statistical importance.]

[A similar argument, applied to the proof of Theorem 4.1 below, shows that the conclusion of Theorem 4.1 holds for the estimate Y_n defined (in a similar modification of the definition of Section 4) as a value of d in the sphere of radius $h_1(n)$ about ϕ_n, which maximizes the integral (4.3). Again as before, it is enough to maximize to within $1 - \ell_n$, $\ell_n \to 0$.]

It follows from (3.25) and the fact that Θ is open that, for large enough n, B_n is always in Θ. Since

$$P_{\theta_0}\{B_n(X(n)) = Z_n(X(n))\} \to 1 ,$$

we have

$$P_{\theta_0}\{Z_n(X(n)) \varepsilon \Theta\} \to 1 .$$

Suppose that, in the argument at the beginning of Section 3(9), ϕ_n had been such that

$$|\phi_n(X_n) - \theta_0| = 0_p(\frac{1}{k^*(n)}) ,$$

where

$$\frac{k^*(n)}{k(n)} \to 0 \ , \quad k^*(n) \to \infty \ .$$

It is then easy to verify that one could obtain Z_n by maximizing the integral (3.3) (over $R_n(d)$) with respect to d in the closed sphere centered at the estimator $\phi_n(X_n)$ and of radius $h_1^*(n) = \frac{1}{k^*(n)}$. The proof is the same as in the previous case.

In [5] a simple example is given where the maximum likelihood estimator is not even consistent. In [15], Section 10, the present authors show that, in the above example, if one limits the domain of θ, with respect to which one maximizes, to a suitable neighborhood of ϕ_n, the m.ℓ. estimator is not only consistent, but equal to the m.p. estimator and efficient.

CHAPTER 4: MAXIMUM PROBABILITY ESTIMATORS WITH A GENERAL LOSS FUNCTION

1) As we remarked in Section 3, the m.p. estimator Z_n with respect to R is really an estimator with respect to a special "loss function," one which assigns loss zero when $k(n)(Z_n-\theta_0)$ is in R (in the limit, as $n \to \infty$), and assigns a positive loss, say one, when $k(n)(Z_n-\theta_0)$ is not in R. We now obtain the m.p. estimator with respect to a general loss function. We postulate the conditions of Section 3.1 up to the introduction of R.

Let $L_n(.,.)$ be a non-negative loss function, i.e., when the value of the estimator is z and the value of the parameter is θ, the loss to the statistician is $L_n(z,\theta)$. The choice of L_n depends upon the statistician, who will take into account the problem and his position (i.e., financial, moral, or whatever is appropriate to the situation). It obviously cannot be prescribed by the mathematician, and the theorem we shall prove is indifferent to the choice of L_n, except insofar as our methods require certain regularity conditions on L_n, conditions which will certainly be satisfied in most practical problems. Any choice of estimator implies, on the part of the statistician, an implicit choice of a class of loss functions with respect to which the estimator is efficient. It may be a problem of the greatest difficulty to determine this class, and it is certainly much more logical and sensible first to choose the loss function explicitly.

We assume that, for every n, $L_n(z,\theta)$ is Borel measurable in both variables jointly, and is Borel measurable in one variable when the other variable is fixed. Since any estimator, say T_n, is a Borel measurable function of its argument (the complex of observations), it follows that $L_n(T_n(x), \theta)$ is a Borel measurable function of x and θ jointly, and a Borel measurable function of x or θ when the other variable is fixed.

For any $y > 0$ define

$$s_n^*(y) = \sup L_n(z,\theta) ,$$

the supremum being taken over all z and θ such that $|z-\theta| \leq y$. Let $\{k_1(n), k_2(n)\}$ be a sequence of pairs of positive numbers such that, as $n \to \infty$,

$$(4.1) \qquad k_2(n) \to \infty, \frac{k_2(n)}{k_1(n)} \to 0, \frac{k_1(n)}{k(n)} \to 0 .$$

Write for brevity

$$h_1(n) = \frac{k_1(n)}{k(n)} , \; h_2(n) = \frac{k_2(n)}{k(n)} ,$$

and

$$(4.2) \qquad s(n) = s_n^*(h_2(n)) .$$

We assume that $s(n) < \infty$ for all n. (This is, among other things, a restriction on $k_2(n)$. It would be enough to require that $s(n) < \infty$ for large n.)

The m.p. estimator Y_n with respect to the (loss) function L_n is a function of $X(n)$ which is that value of d which maximizes the integral

$$(4.3) \qquad \int_{d-h_2(n)}^{d+h_2(n)} [s(n) - L_n(d,\theta)]K_n(X(n)|\theta)d\theta .$$

Our definition assumes that Y_n exists and is unique, and we shall proceed under these assumptions. First, though, we will discuss them very briefly.

Let $d_1, d_2, \ldots$ be a sequence of numbers such that the integral (4.3), evaluated at $d = d_1, d_2, \ldots$ approaches its supremum (not ∞), and such that d_2 converges (not to $+\infty$ or $-\infty$). In this case, if Y_n does not exist, it is sufficient to maximize the integral (4.3) to within ℓ_n ($\ell_n > 0$, $\ell_n \to 0$), exactly as in (3.4). Suppose that there exist at least two sequences of d's as at the beginning of this paragraph, each with a different limit. Then, for almost all reasonable problems which will occur in actual statistical practice, one of these sequences will have θ_0, the true value of θ, as its limit. When this is the case we proceed as in the paragraph which follows (3.4), and maximize the integral (4.3) with respect to d in the closed sphere (there described) centered at ϕ_n and of radius $h_1(n)$. Finally, even when the problem is such that no sequence of d's converges to the true value of θ (for some or all θ in Θ), it may perhaps be possible first to estimate the true value of θ for some R as in Section 3, and then to maximize the integral (4.3) with respect to d in a set determined by this last estimator. Thus the assumptions, which we henceforth make, that Y_n exists and is unique, are not very restrictive, and are really made chiefly for convenience of exposition.

We shall say that a sequence $\{\theta_n, n = 1,2,\ldots\}$ of real numbers is in $H(k_1)$ if $k(n)|\theta_n - \theta_0| \leqq k_1(n)$ for $n = 1,2,\ldots$. The definition thus depends implicitly on θ_0 and $k(n)$. We will prove

<u>Theorem</u> 4.1. Suppose that the estimator Y_n satisfies the following three conditions, for $\{\theta_n\}$ in $H(k_1)$:

$$\lim_{n\to\infty} E_{\theta_n}\{L_n(Y_n,\theta_n)\} = \beta(\theta_0), \text{ say,} \tag{4.4}$$

$$\lim_{n\to\infty} [s(n)P_{\theta_n}\{|k(n)(Y_n-\theta_n)| > k_2(n)\}] = 0 , \tag{4.5}$$

and

$$\lim_{n\to\infty} \int_{D_n(\theta_n)} L_n(Y_n(x),\theta_n)K_n(x|\theta_n)d\mu_n(x) = 0 , \tag{4.6}$$

where

$$D_n(\theta_n) = \{x \,\big|\, |k(n)(Y_n-\theta_n)| > k_2(n)\} . \tag{4.7}$$

Let T_n be any estimator for which the following two conditions hold, for $\{\theta_n\}$ in $H(k_1)$:

$$\lim_{n\to\infty} [E_{\theta_n}\{L_n(T_n,\theta_n)\} - E_{\theta_0}\{L_n(T_n,\theta_0)\}] = 0 \tag{4.8}$$

and

$$\lim_{n\to\infty} [s(n)P_{\theta_n}\{|k(n)(T_n-\theta_n)| > k_2(n)\}] = 0 . \tag{4.9}$$

Then

$$\beta(\theta_0) \leq \underline{\lim}_{n\to\infty} E_{\theta_0}\{L_n(T_n,\theta_0)\} , \tag{4.10}$$

so that Y_n is asymptotically efficient in this sense.

2) For the reasons given in Section 3, the relevant hypotheses of Theorem 4.1 are also formulated in terms of the behavior of the maximum probability estimator, and not in terms of $K_n(\cdot|\cdot)$. Of course the hypotheses are also restrictions on the loss function L_n. In most reasonable and practical statistical problems conditions (4.4)-(4.6) will be satisfied for suitably chosen $k_1(\cdot)$ and $k_2(\cdot)$.

It is difficult to imagine a statistician's using an estimator which does not satisfy (4.8). The reasons are essentially the same as those which were given for the analogous condition (3.7). Condition (4.9) is also a condition on $s(n)$. While it is a not unreasonable condition and will be satisfied in many cases, the argument for it is not at all as compelling as that for (4.8). Condition (4.9) may well be needed only because of our particular method of proof and/or the desire to include all loss functions which are bounded below. Theorem 4.2 below already shows that (4.9) is not needed for many loss functions natural in practical applications.

<u>Theorem</u> 4.2 If, for all n sufficiently large, $L_n(z,\theta)$ is a monotonically non-decreasing function of $|z-\theta|$, Theorem (4.1) holds even without the condition (4.9).

<u>Theorem</u> 4.3 If, for all n sufficiently large, $L_n(z,\theta) = s(n)$ for $|z-\theta| > h_2(n)$, Theorem 4.1 holds even without the condition (4.9).

One more remark before we proceed to the proofs. Suppose $\underline{\lim}\ s(n) = 0$. It follows from (4.5) that then $\underline{\lim}\ E_{\theta_0}\{L_n(Y_n,\theta_0)\} = 0$. Hence, from (4.4), $\beta(\theta_0) = 0$, and Theorems 4.1-4.3 hold.

We may therefore assume that $\underline{\lim}\ s(n) > 0$. If, for all n sufficiently large, $0 < a_1 < s(n) < a_2 < \infty$, conditions (4.5) and (4.9) take a particularly simple form.

3) Proof of Theorem 4.1. Suppose that the theorem is not true and that

(4.11) $$\beta(\theta_0) - \underline{\lim}_{n\to\infty} E_{\theta_0}\{L_n(T_n,\theta_0)\} = 4\gamma > 0 \ .$$

Define $L_n^*(d,\theta)$ as follows:

$$L_n^*(d,\theta) = L_n(d,\theta) \quad \text{when} \quad k(n)|\theta-d| \leq k_2(n) \ ,$$

$$L_n^*(d,\theta) = s(n) \quad \text{when} \quad k(n)|\theta-d| > k_2(n) \ .$$

Let $B_n^*(X(n))$ be an estimator which minimizes, with respect to B_n, the integral

(4.12) $$\int_{A_n^*} \int L_n^*(B_n(x),\theta)K_n(x|\theta)d\mu_n(x)d\theta \ ,$$

where $A_n^* = \{\theta \mid k(n)\ |\theta-\theta_0| \leq k_1(n)\}$. (If a minimizing estimator does not exist it is sufficient to use a B_n^* which minimizes the integral to within $\ell_n > 0$, where $\ell_n \to 0$. We have used this device earlier.) Inverting the order of integration in (4.12) this means that, for each x, the integral

(4.13) $$\int_{A_n^*} L_n^*(B_n(x),\theta)K_n(x|\theta)d\theta$$

is a minimum. (An x-set of μ_n measure zero may be excepted.) From the definitions of L_n^* and A_n^* it follows that, whenever

(4.14) $$|Y_n(x) - \theta_0| \leq h_1(n) - h_2(n) ,$$

we may set $B_n^*(x) = Y_n(x)$, and we will actually do so. The inequalities

(4.15) $$|\theta-\theta_0| \leq h_1(n) - 2h_2(n)$$

and

(4.16) $$|Y_n(x) - \theta| \leq h_2(n)$$

imply (4.14). Now consider the integral

(4.17) $$[2h_1(n)]^{-1} \int E_\theta\{L_n^*(Y_n,\theta)\}d\theta$$

over the set $D_n' = \{\theta \mid (h_1(n)-2h_2(n)) < |\theta-\theta_0| \leq h_1(n)\}$. It follows from (4.4), (4.5), and (4.6) that

(4.18) $$\lim_{n\to\infty} E_\theta\{L_n^*(Y_n,\theta)\} = \beta(\theta_0)$$

for θ in A_n^*. Hence, from (4.1), we have

(4.19) $$\lim_{n\to\infty} [2h_1(n)]^{-1} \int_{D_n'} E_\theta\{L_n^*(Y_n,\theta)\}d\theta = 0 .$$

From (4.15), (4.16), (4.19), and (4.5) we obtain that, for sufficiently large,

(4.20)

$$[2h_1(n)]^{-1} \int_{A_n^*} E_\theta\{L_n^*(Y_n,\theta)\}d\theta < [2h_1(n)]^{-1} \int_{A_n^*} E_\theta\{L_n^*(B_n^*,\theta)\}d\theta + \gamma$$

$$\leq [2h_1(n)]^{-1} \int_{A_n^*} E_\theta\{L_n^*(T_n,\theta)\}d\theta + \gamma .$$

The last member of (4.20) is not greater than

$$(4.21) \qquad [2h_1(n)]^{-1} \int_{A_n^*} E_\theta\{L_n(T_n,\theta)\}d\theta + \gamma$$

$$+ s(n)[2h_1(n)]^{-1} \int_{A_n^*} P_\theta\{k(n)|T_n-\theta| > k_2(n)\}d\theta .$$

From (4.20), (4.21), (4.6), and (4.9) we obtain that, for n sufficiently large,

(4.22)

$$[2h_1(n)]^{-1} \int_{A_n^*} E_\theta\{L_n(Y_n,\theta)\}d\theta < [2h_1(n)]^{-1} \int_{A_n^*} E_\theta\{L_n(T_n,\theta)\}d\theta + 2\gamma .$$

From (4.22), (4.4), and (4.8) we obtain

$$(4.23) \qquad \beta(\theta_0) < \underline{\lim}_{n\to\infty} E\{L_n(T_n,\theta_0)\} + 3\gamma .$$

This contradicts (4.11) and proves Theorem 4.1.

Proof of Theorem 4.2 In the proof of Theorem 4.1 the condition (4.9) was used only to deduce (4.22) from (4.21). If L_n is a monotonically non-decreasing function of $|z-\theta|$, (4.22) follows directly from (4.20) and (4.6), without the intervention of (4.21). This proves Theorem 4.2.

Proof of Theorem 4.3 Suppose that, for all n sufficiently large, $L_n(z,\theta) = s(n)$ for $|z-\theta| > h_2(n)$. Then, in (4.20), we may replace L_n^* by L_n. This again gives us (4.22) without the intervention of (4.21) and hence without the use of (4.9). This proves Theorem 4.3.

Consider the following illustration. Let μ_n be n-dimensional Lebesgue measure, $X(n) = (X_1,\ldots,X_n)$, $m = 1$, Θ the entire real line, and $K_n(x_1,\ldots,x_n|\theta) = (2\pi)^{-n/2}\exp\{-\frac{1}{2}\sum_1^n (x_i-\theta)^2\}$. Let $L_n(z,\theta) = n(z-\theta)^2$, $k(n) = n^{1/2}$, $k_1(n) = n^{1/3}$, $k_2(n) = n^{1/4}$. Then $s(n) = n^{1/2}$. We conclude that $\overline{X}(n) = n^{-1}\sum_1^n X_i$ is the maximum probability estimator Y_n. Hence any estimator T_n which satisfies (4.8) also satisfies, by Theorem 4.2, the inequality

$$(4.24) \qquad \underline{\lim}_{n\to\infty} E_\theta\{(T_n-\theta)^2\} \geq 1$$

for any θ. The estimator T_n need not be unbiased or asymptotically normally distributed. Of course, if T_n is asymptotically normally distributed about θ it would be more intelligent to concern one's self with the variance of the limiting distribution. This can be done using an R which is an interval centered at the origin. If the limiting distribution of $(T_n-\theta)$ is not normal the second moment may well not be the appropriate measure of loss. Of course, this does not affect the validity of (4.24) or its value as an illustration.

The multi-dimensional $(m > 1)$ analogue of Theorem 4.1 enables us to extend the result described at the end of Secti 3.6 to unbounded R.

CHAPTER 5: ASYMPTOTIC BEHAVIOR OF THE LIKELIHOOD FUNCTION. ASYMPTOTICALLY SUFFICIENT STATISTICS.

5(1) In Chapter 3 we explained why the conditions of our theorems are put in terms of conditions on the behavior of the m.p. estimator. To repeat briefly, the conditions on the m.p. estimator are relatively easy to verify, while conditions on $K_n(\cdot|\cdot)$, if they were to be applicable in the desired generality to most important cases, would have to be of great complexity and difficult to verify.

In this section we give several different sets of conditions on K_n. Each set covers a large class of densities f (with respect to Lebesgue measure). We will always be dealing with independent identically distributed chance variables $X_1,\dots,X_n$. The disparities among the different conditions point up the hopelessness of subsuming these and many others in one manageable set of conditions. Our conditions are formulated only for convenience, are much stronger than actually necessary, and can easily be weakened. For more examples see, inter alia, those of [12], [15], and those of this monograph. The regular case is deliberately omitted from this section, because almost all the literature is devoted to it, and is postponed until Chapter 6.

For each set of conditions we will determine an m.p. estimator Z_n with respect to $R = (-r,r)$, and show (if it is necessary; usually it will be obvious) that conditions (3.5) and (3.6) are satisfied. In each case the asymptotic behavior of the likelihood function in a neighborhood of θ_0 will be shown to be such that, for the purpose of finding Z_n with respect to R, the likelihood function depends solely on a statistic which we therefore describe as "asymptotically sufficient with respect to R." This (new) concept of asymptotic sufficiency differs, inter alia, from other definitions given in the literature in that it depends on R (or the loss function L_n). Perhaps the essential reason that other definitions of asymptotic sufficiency do not depend on R is that they are oriented towards the regular case and asymptotically normal estimators. (For the same reason the classical theory does not employ R, which was introduced only by the present theory.) For the latter estimators we have already seen in Section 3(6) that, essentially, we can limit ourselves to an R which is any interval centered at the origin. We have chosen this class of R's for this chapter because it is important, relatively easy to compute with, and some class had to be chosen for definiteness of illustration.

In this section we will always obtain Z_n and Y_n by maximizing in the sphere centered at ϕ_n and of radius $h_1(n)$, as described in Chapter 3, where the validity of this procedure was proved. The estimator ϕ_n is usually easy to obtain. Often the m.ℓ. estimator is such an estimator, but is not necessarily efficient.

5(2) Let Θ be the real line and μ_n be n-dimensional Lebesgue measure. Let $f(x|\theta)$ be continuous in (x,θ) for all (x,θ) such that $x > \theta$. Let $f(x|\theta) = 0$, $x < \theta$, and $f(x|\theta) > 0$, $x \geqq \theta$. Suppose $f(\theta+|\theta) = h(\theta) > 0$. We assume that $\frac{\partial f(x|\theta)}{\partial\theta}$ exists and is continuous in (x,θ) for all (x,θ) such that $x > \theta$. Hence

$$(5.1) \qquad \int_{\theta}^{\infty} \frac{\partial f(x|\theta)}{\partial\theta}\, dx = h(\theta) .$$

We also assume that $\frac{\partial^2 \log f(x|\theta)}{\partial\theta^2}$ exists for $x > \theta$ and is bounded in absolute value by D, say, and that, for all θ in Θ,

$$(5.2) \qquad \int_\theta^\infty \left[\frac{\partial \log f(x|\theta)}{\partial\theta}\right]^2 f(x|\theta)\, dx < \infty .$$

An example of such a density is the following:

$$f(x|\theta) = e^{-(x-\theta)}, \quad x \geq \theta, \quad \text{and zero otherwise.}$$

Define $w_n = \min(X_1,\ldots,X_n)$. Let $\phi_n = w_n$, $k(n) = n$, $k_1(n) = n^{1/4}$. We have

$$(5.3) \qquad P_{\theta_0}\{0 < w_n - \theta_0 \leqq h_1(n)\} \to 1 .$$

Hence, for all θ such that

$$(5.4) \qquad w_n - 2h_1(n) \leqq \theta < w_n ,$$

we have

$$(5.5) \qquad P_{\theta_0}\Big\{\Big|\log K_n(X(n)|\theta) - \log K_n(X(n)|\theta_0) - (\theta-\theta_0)\sum_1^n \frac{\partial \log f(X_i|\theta_0)}{\partial\theta}\Big| < \frac{n(\theta-\theta_0)^2 D}{2}\Big\} \to 1 .$$

Since $w_n - \theta_0 = O_p(\frac{1}{n})$, it follows from (5.4) that

$$(5.6) \qquad n(\theta-\theta_0)^2 = O_p(n^{-1/2}) .$$

Define

$$M_1 = n^{-1} \log K_n(X(n)|\theta_0)$$

$$M_2(\theta) = n^{-1} \log K_n(X(n)|\theta) .$$

It follows from the central limit theorem that

$$(5.7) \qquad n^{-1} \sum_1^n \frac{\partial \log f(X_i|\theta_0)}{\partial\theta} = h(\theta_0) + 0_p(n^{-1/2}) .$$

Hence, for any $\varepsilon > 0$ there exists a constant $K_\varepsilon > 0$ such that, when θ satisfies (5.4) and n is sufficiently large,

(5.8)

$$P_{\theta_0}\{\exp(nM_2(\theta)) = \exp(n[M_1 + (\theta-\theta_0)(h(\theta_0) + c_\varepsilon(\theta)n^{-1/2})])\} > 1 - \varepsilon$$

and $|c_\varepsilon(\theta)| < K_\varepsilon$. When $\theta \geqq w_n$, $K_n(X(n)|\theta) = 0$.

Let us now maximize, with respect to d in the closed interval centered at w_n and of half-length $n^{-3/4}$, the integral

$$(5.9) \qquad \int_{d-\frac{r}{n}}^{d+\frac{r}{n}} K_n(X(n)|\theta)d\theta .$$

Since $K_n(X(n)|\theta) = 0$ when $\theta \geqq w_n$, this amounts to restricting d to the closed interval

$$(5.10) \qquad [w_n - n^{-3/4}, \; w_n - \frac{r}{n}] .$$

The interval (5.10) is contained within the interval (5.4). Bearing (5.8) in mind, suppose that we maximize instead the integral

$$(5.11)\qquad \int_{d-\frac{r}{n}}^{d+\frac{r}{n}} \exp\{n[M_1 + (\theta-\theta_0)(h(\theta_0) + c_\varepsilon(\theta)n^{-1/2})]\}d\theta$$

with respect to d in (5.10). This will give us the value of Z_n with probability greater than $1-\varepsilon$ for large n. Conditional upon the event in (5.8), the maximum of the integral in (5.11) is $\geq \exp\{-K_\varepsilon n^{-1/2}\}$ multiplied by the maximum of the integral (5.9). Since ε was arbitrary, it follows from the conclusion which follows (3.4) that the value of d which maximizes (5.11) is asymptotically equivalent to the m.p. estimator Z_n.

As for the integral (5.11), when n is large $h(\theta_0) + c_\varepsilon(\theta)n^{-1/2} > 0$, so that then the integral (5.11) is monotonically increasing with θ. Hence, from (5.10), its maximizing value depends only on w_n, which is therefore asymptotically sufficient for R. The statistician need not know the entire sample $X(n)$; it is sufficient that he know only w_n. The estimator w_n is, of course, an m.p. estimator.

We verify easily that Z_n satisfies the regularity conditions of Theorem 3.1.

Now consider a general loss function L_n. Since $w_n = \theta_0 + 0_p(\frac{1}{n})$ we have

$$(5.12)\qquad h(w_n) = h(\theta_0) + 0_p(\tfrac{1}{n}) \; .$$

Define

$$(5.13)\qquad M_2'(\theta) = M_1 + (\theta-\theta_0)h(w_n), \qquad \theta < w_n \; .$$

Hence, for all θ in the interval (5.4), we have from (5.8) that, for large n,

(5.14) $P_{\theta_0}\{\exp(nM_2(\theta)) = \exp(nM_2'(\theta)) \cdot \exp(c_\varepsilon'(\theta)n^{-1/4})\} > 1 - \varepsilon$,

where $|c_\varepsilon'(\theta)| < K_\varepsilon$. Define

$$\chi(\theta,w_n) = 1, \quad \theta < w_n$$

$$\chi(\theta,w_n) = 0, \quad \theta \geqq w_n .$$

Let d_1 be such that

$$c_1 = \int_{d_1-h_2(n)}^{d_1+h_2(n)} \chi(\theta,w_n)[s(n) - L_n(d_1,\theta)]\exp\{nM_2(\theta)\}d\theta$$

is a maximum, d_2 be such that

$$c_2 = \int_{d_2-h_2(n)}^{d_2+h_2(n)} \chi(\theta,w_n)[s(n) - L_n(d_2,\theta)]\exp\{nM_2'(\theta)\}d\theta$$

is a maximum, and let

$$c_3 = \int_{d_1-h_2(n)}^{d_1+h_2(n)} \chi(\theta,w_n)[s(n) - L_n(d_1,\theta)]\exp\{nM_2'(\theta)\}d\theta .$$

Then we have, conditional upon the event in (5.14), that

(5.15) $c_2 \geqq c_3 \geqq c_1 (1 - K_\varepsilon n^{-1/4})$.

Since ε was arbitrary we have proved that an m.p. estimator Y_n with respect to L_n can be obtained by maximizing

(5.16) $$\int_{d-h_2(n)}^{d+h_2(n)} \chi(\theta,w_n)[s(n) - L_n(d,\theta)]\exp\{n\theta h(w_n)\}d\theta$$

(see c_2) with respect to d in the interval

$$(5.17) \qquad [w_n - h_1(n), w_n + h_1(n)] .$$

It remains to choose k_2 and verify the conditions on Y_n of Theorem 4.1. This surely requires information about the function L_n. From (5.16) and (5.17) it follows that w_n is an asymptotically sufficient statistic with respect to L_n. The statistician need not know the entire sample $X(n)$; it is sufficient that he know w_n only.

As we have seen, in order that the statistician find Z_n with respect to R, it is not necessary that he know the function $h(\cdot)$. In general, the determination of Y_n from (5.16) and (5.17) (or by any other method) will require knowledge of $h(\cdot)$.

5(3) Suppose now that always $\theta < A(\theta)$ (a known function of θ), $\frac{dA(\theta)}{d\theta}$ negative and continuous, and that

$$f(x|\theta) = 0, \quad x < \theta \quad \text{or} \quad x > A(\theta) .$$

Suppose also that

$$f(\theta+ \mid \theta) = h(\theta) > 0 ,$$

$$f(A(\theta) - \mid \theta) = g(\theta) > 0 ,$$

where h and g are continuous functions. Define $v_n = \max(X_1,\ldots,X_n)$. Then $K_n(X(n)|\theta)$ can be positive only when

$$(5.18) \qquad \theta < \min \{w_n, A^{-1}(v_n)\} = G(w_n,v_n), \quad \text{say.}$$

We leave it to the reader to impose regularity conditions similar to those of Section 5(2) so that $k(n) = n$ and, with $k_1(n) = n^{1/4}$, (5.5) should hold in the interval

(5.19) $$G(w_n, v_n) - 2h_1(n) < \theta < G(w_n, v_n) .$$

Consider the sign of

(5.20) $$E_{\theta_0}\{\frac{\partial \log f(X_1|\theta_0)}{\partial\theta}\} = C(\theta_0), \quad \text{say.}$$

Differentiating both sides of

(5.21) $$1 \equiv \int_{\theta}^{A(\theta)} f(x|\theta)dx$$

with respect to θ, we obtain that

(5.22) $$C(\theta_0) = h(\theta_0) - A'(\theta_0)g(\theta_0) > 0 ,$$

since $A'(\theta_0) < 0$. Thus, by an argument like that of the previous section, we conclude that $Z_n = G(w_n, v_n) - \frac{r}{n}$. The pair (w_n, v_n) is asymptotically sufficient with respect to R. It is easy to show that (3.5) and (3.6) are satisfied.

5(4) This example is the same as that of the preceding section, except that now always $\frac{dA(\theta)}{d\theta} > 0$.

We know that, with probability one,

(5.23) $$\theta_0 < w_n < v_n < A(\theta_0) .$$

Hence, w.p.1,

(5.24) $A^{-1}(v_n) < \theta_0 < w_n$.

Outside the θ-interval

(5.25) $[A^{-1}(v_n), w_n]$

$K_n(X(n)|\theta)$ is zero. Inside the interval we represent $\log K_n(X(n)|\theta)$ in the form (5.5). Again, as in the previous section, the crucial question is the sign of $C(\theta_0)$ whose value is given by the second member of (5.22).

Case 1: $C(\theta_0) < 0$. Then $\log K_n(X(n)|\theta)$ is essentially decreasing in the crucial interval, and $Z_n = A^{-1}(v_n) + \frac{r}{n}$.

Case 2: $C(\theta_0) > 0$. Then $\log K_n(X(n)|\theta)$ is essentially increasing in the crucial interval and $Z_n = w_n - \frac{r}{n}$.

Case 3: $C(\theta_0) = 0$. Then $\log K_n(X(n)|\theta)$ is essentially constant in the crucial interval (essentially constant means modulo the argument based on (5.5)), and we obtain the following conclusion: Z_n may be any point in the interval

$$[m^*, m^{**}] ,$$

where

$$m^* = \min\{w_n - \frac{r}{n} , A^{-1}(v_n) + \frac{r}{n}\}$$

$$m^{**} = \max\{w_n - \frac{r}{n} , A^{-1}(v_n) + \frac{r}{n}\}$$

provided that it (Z_n) is Borel measurable. The pair (w_n, v_n) is asymptotically sufficient with respect to R. The regularity conditions are not difficult to verify.

For other loss functions one can proceed as in (5.16).

5(4) We now briefly consider a two-dimensional generalization of the density of Section 5(2). (For a description of the m.p. estimator when $m > 1$ see [13], Section 3.) Let Θ be the real plane, $\theta = (\theta_1,\theta_2)$, $\theta_0 = (\theta_{10},\theta_{20})$, $X_i = (X_{1i},X_{2i})$, $w_{1n} = \min(X_{11},\dots,X_{1n})$, $w_{2n} = \min(X_{21},\dots,X_{2n})$. Assume such regularity conditions that the two-dimensional analogue of the argument of Section 5(2) applies and that

$$E_{\theta_0}\left\{\frac{\partial \log f(X_1|\theta)}{\partial\theta_1}\bigg|_{\theta=\theta_0}\right\} > 0$$

and

$$E_{\theta_0}\left\{\frac{\partial \log f(X_1|\theta)}{\partial\theta_2}\bigg|_{\theta=\theta_0}\right\} > 0 .$$

An argument analogous to that of Section 5(2) leads to the following conclusion: In the intersection of the set $\{\theta_1 \leqq w_{1n},\ \theta_2 \leqq w_{2n}\}$ with a suitable neighborhood of $w_n = (w_{1n},w_{2n})$, the likelihood function (of θ) $K_n(X(n)|\theta)$ is monotonically increasing. This happens with probability approaching one.

Our generalization of the region R is the following:

$$R' = \{(\theta_1,\theta_2) \,\big|\, |\theta_1| \leqq r_1,\ |\theta_2| \leqq r_2\} .$$

For this R' the m.p. estimator Z_n is given by

$$Z_n = \left(w_{1n} - \frac{r_1}{n},\ w_{2n} - \frac{r_2}{n}\right) .$$

Other results of this section can be similarly extended.

5(5) Let Θ be the real line and

$$f(x|\theta) = \frac{1}{2} \exp\{-|x-\theta|\} .$$

Suppose $n = 2m + 1$, m an integer. Let ϕ_n be the median of the n (independent) X's, $k(n) = n^{1/2}$, and, say, $k_1(n) = n^{1/32}$, $k_2(n) = n^{1/64}$. As earlier in this chapter, let $R = \{\theta|\ |\theta| \leqq r\}$. The m.p. estimator Z_n maximizes

$$\int_{d-rn^{-1/2}}^{d+rn^{-1/2}} K_n(X(n)|\theta)d\theta .$$

The likelihood function $K_n(X(n)|\theta)$ is always continuous in θ, strictly increasing when $\theta < \phi_n$ and strictly decreasing when $\theta > \phi_n$. Consequently Z_n satisfies

$$K_n(X(n)|Z_n - rn^{-1/2}) = K_n(X(n)|Z_n + rn^{-1/2}) . \tag{5.26}$$

In [12] it is essentially proved that, when Z_n satisfies (5.26), $\sqrt{n}\,(Z_n-\theta_0)$ and $\sqrt{n}\,(\phi_n-\theta_0)$ have the same limiting distribution. Hence we may set $Z_n = \phi_n$. It is not difficult to verify the regularity conditions of Theorem 3.1.

CHAPTER 6: EFFICIENCY OF MAXIMUM LIKELIHOOD ESTIMATORS

The classical regular case is the one to which most of the literature is devoted and the only one for which the literature gives an asymptotically efficient estimator, namely, the maximum likelihood estimator. In the first section of this chapter we make rigorous the heuristic argument of Section 3(5). In the second section we discuss an interesting non-regular case for which the m.ℓ. estimator is efficient.

Throughout this chapter we assume that $R = (-r,r)$ is any interval centered at the origin, and that $X_1, X_2, \ldots$ are independent chance variables with the common density $f(\cdot|\theta)$. The m.ℓ. estimator $\hat{\theta}_n$ will be assumed to satisfy the likelihood equation

$$\sum_{i=1}^{n} \frac{\partial \log f(X_i|\theta)}{\partial\theta} = 0 .$$

6(1) We shall use a small modification of the regularity conditions of [6]. (See the second paragraph of Section 3(3). The reader who so desires may omit reading the regularity conditions without losing the argument.) Thus we assume:

(6.1) The function $\theta \to P_\theta$ is continuous on Θ with respect to the metric defined by the distance function

$$d(P,Q) = \sup_{A} |P\{A\} - Q\{A\}| ,$$

the supremum taken over all Borel sets.

(6.2) For each real x, the function $\theta \to \log f(x|\theta)$ is continuous on $\overline{\Theta}$, the closure of Θ.

(6.3) For every $\theta \in \Theta$ and every compact set $K \subset \Theta$:

a) $\sup_{\tau \varepsilon K} E_\tau [\log f(X|\tau)]^2 < \infty$

b) $\log f(\cdot|\theta)$ is uniformly integrable with respect to E_τ, $\tau \varepsilon K$.

(6.4) For each x, $\log f(x|\theta)$ is twice differentiable with respect to θ (in Θ). Also, for all θ in Θ,

$$E_\theta \left[\frac{\partial \log f(X|\theta)}{\partial \theta} \right] = 0$$

(6.5) For every compact set $K \subset \Theta$,

a) $$\inf_{\theta \varepsilon K} E_\theta \left[\frac{\partial \log f(X|\theta)}{\partial \theta} \right]^2 > 0$$

b) $$\inf_{\theta \varepsilon K} E_\theta \left[\frac{\partial^2 \log f(X|\theta)}{\partial \theta^2} - E_\theta \left(\frac{\partial^2 \log f(X|\theta)}{\partial \theta^2} \right) \right]^2 > 0$$

c) $$\sup_{\theta \varepsilon K} E_\theta \left[\frac{\partial^2 \log f(X|\theta)}{\partial \theta^2} \right] < 0$$

(6.6) For every compact set $K \subset \Theta$,

a) $$\sup_{\theta \varepsilon K} E_\theta \left[\left| \frac{\partial \log f(X|\theta)}{\partial \theta} \right|^3 \right] < \infty$$

b) $$\sup_{\theta \varepsilon K} E_\theta \left[\left| \frac{\partial^2 \log f(X|\theta)}{\partial \theta^2} \right|^3 \right] < \infty$$

(6.7) For every θ in $\overline{\Theta}$ there exists a neighborhood U_θ of θ such that

a) For every neighborhood U of θ such that $U \subset U_\theta$, and every compact set $K \subset \Theta$,

$$\sup_{\theta \varepsilon K} E_\theta \left[\sup_{\tau \varepsilon U} \log f(X|\tau)\right]^2 < \infty$$

b) for every compact set $K \subset \Theta$ the function

$$\sup_{\theta \varepsilon U_\theta} \log f(x|\theta)$$

is uniformly integrable from above with respect to E_τ, $\tau \varepsilon K$.

(6.8) For every $\theta \varepsilon \Theta$ there exists an open neighborhood V_θ of θ and a constant k_θ such that, for all x and all τ_1 and τ_2 in V_θ,

$$\left| \frac{\partial^2 \log f(x|\theta)}{\partial\theta^2}\bigg|_{\tau_1} - \frac{\partial^2 \log f(x|\theta)}{\partial\theta^2}\bigg|_{\tau_2} \right| \leqq k_\theta|\tau_1-\tau_2| .$$

(6.9) For every θ there is a neighborhood U_θ of θ and a constant d_θ such that, for any pair τ_1 and τ_2 in U_θ,

$$|I(\tau_1) - I(\tau_2)| \leqq d_\theta \; |\tau_1-\tau_2| .$$

This completes the regularity conditions. (See the comments at the end of this section.) From these it follows easily that $\sqrt{n}\,(\hat\theta_n-\theta_0)$ is in the limit normally distributed, with mean zero and variance $[I(\theta_0)]^{-1}$. It remains to prove that we may take $Z_n = \hat\theta_n$ and that $\hat\theta_n$ fulfills the regularity conditions (3.5) and (3.6). From these two statements the conclusion (3.18) will follow exactly as in the heuristic argument of Section 3(6), for any competing estimator T_n which satisfies (3.7) and the classical requirement that $\sqrt{n}\,(T_n-\theta_0)$ be asymptotically normally distributed with mean zero.

To make (3.16) rigorous it is sufficient to write

$$(6.10)\qquad \sum_{i=1}^{n} \log f(X_i|\theta) = \sum_{i=1}^{n} \log f(X_i|\hat{\theta}_n) + \frac{n(\theta-\hat{\theta}_n)^2}{2}\left[\frac{1}{n}\sum_{i=1}^{n} \frac{\partial^2 \log f(X_i|\theta)}{\partial\theta^2}\bigg|_{\xi}\right],$$

where ξ lies between θ and $\hat{\theta}_n$. We now restrict θ to lie in the interval A_n of Section 3(4). Let $\varepsilon > 0$ be arbitrary and then h be sufficiently large. We have, for all n large enough,

$$(6.11)\qquad P_{\theta_0}\{\hat{\theta}_n \in A_n\} > 1 - \varepsilon .$$

When $\hat{\theta}_n \in A_n$ we have, from (6.8) and (6.10),

(6.12)

$$\sum_{i=1}^{n} \log f(X_i|\theta) = \sum_{i=1}^{n} \log f(X_i|\hat{\theta}_n) + \frac{n(\theta-\hat{\theta}_n)^2}{2}\left[\frac{1}{n}\sum_{i=1}^{n} \frac{\partial^2 \log f(X_i|\theta)}{\partial\theta^2}\bigg|_{\theta_0}\right] + \frac{n(\theta-\hat{\theta}_n)^2}{2}\, 0\left[\frac{1}{\sqrt{n}}\right].$$

The last term in the right member of (6.12) is $0_p\left[\frac{1}{\sqrt{n}}\right]$, and the quantity in square brackets is $0_p(1)$ and negative with probability approaching one. Thus there exists a constant $c > 0$ such that, for all n sufficiently large, the P_{θ_0}-probability is greater than $1-2\varepsilon$ that $\hat{\theta}_n$ maximizes the right member of (3.4) (with respect to d) with $\ell_n \leq \exp \frac{c}{\sqrt{n}} - 1$. Since ε was arbitrary it follows that Z_n is asymptotically equivalent to $\hat{\theta}_n$, i.e.,

$$(6.13) \qquad P_{\theta_0}\{Z_n = \hat{\theta}_n\} \to 1 .$$

It remains to show that $\hat{\theta}_n$ satisfies (3.5) and (3.6). Let the U_{θ_0} of (6.9) be contained in a finite interval; obviously, this can always be done. It is proved in [6] that there exists a constant c^* such that, for any θ in U_{θ_0} and any y,

(6.14)

$$\left| P_\theta\{\sqrt{n}(\hat{\theta}_n-\theta) < y\} - \sqrt{\frac{1}{2\pi}} \int_{-\infty}^{y[I(\theta)]^{1/2}} \exp\{-\frac{z^2}{2}\}\, dz \right| \leq \frac{c^*\sqrt{\log n}}{\sqrt{n}} .$$

Now let τ_1 and τ_2 be any two points in A_n. It follows from (6.9) and (6.14) that

$$(6.15) \quad |P_{\tau_1}\{\sqrt{n}(\hat{\theta}_n-\tau_1) < y\} - P_{\tau_2}\{\sqrt{n}(\hat{\theta}_n-\tau_2) < y\}| = 0\left[\sqrt{\frac{\log n}{n}}\right].$$

Since $R = (-r,r)$, (3.5) (for the present problem) follows at once from (6.15), setting $y = r$ and $y = -r$.

To prove (3.6), choose ε and δ, and then h so large that

$$(6.16) \qquad \sqrt{\frac{1}{2\pi}} \int_{-\infty}^{\delta h[I(\theta_0)]^{1/2}} \exp\{-\frac{z^2}{2}\}\, dz > 1 - \frac{\varepsilon}{4} .$$

From (6.15) for $y = \pm\delta h$ one easily obtains that (3.6) is satisfied. This completes our proof of the classical Fisher inequality (3.18).

The postulated regularity conditions are stronger than necessary and can be weakened. The result of [6], (6.14), which follows from them and which was used by us, is valid for any compact subset of Θ, whereas we need it only for a neighborhood which can be very small.

6(2) Let $f(\cdot)$ be a uniformly continuous density which vanishes on $(-\infty,0]$, and let $f(x|\theta) = f(x-\theta)$, $\Theta = (-\infty,\infty)$. Define α_n by $2\alpha_n^2 = \alpha n \log n$, where α is defined in (6.18) below. First we write the following conditions:

(6.17) $\{x \mid f(x) > 0\} = (0,b), \quad 0 < b \leq \infty .$

(6.18) f is continuously differentiable on $(0,b)$ with derivative f', $\alpha = \lim f'(x)$ as $x \downarrow 0$ exists, and $0 < \alpha < \infty$. On every compact subinterval of $(0,b)$ f' is absolutely continuous with derivative f'', and $\lim xf''(x) = 0$ as $x \downarrow 0$.

(6.19) With $g = \log f$, for every $\delta > 0$,

$$\int_\delta^b (g'(x))\, f(x)dx < \infty .$$

(6.20) For every a, $0 < a < b$, there exist $\eta > 0$ and $0 < \delta < \min(x,\eta)$ such that

$$\int_a^{b-\eta} \sup_{|t|\leq\delta} |g''(x-t)|\, f(x)dx < \infty ,$$

where $b-\eta = \infty$ if $b = \infty$.

(6.21) With $b < \infty, -\infty \leq \limsup g''(x) < \infty$ as $x \uparrow b$, and there exist $\Pi > 0$, $\beta < 1$, and an increasing function h on

$(0, \eta\beta^{-1})$ for which $g''(x) \geqq h(b-x)$, $b - \eta < x < b$, and

$$\int_{b-\eta}^{b} h(\beta(b-x))f(x)dx > -\infty .$$

Let $\{\hat{\theta}_n\}$ be a consistent sequence. It is then proved in [22] that $\alpha_n(\hat{\theta}_n - \theta_0)$ has a limiting distribution which is normal with mean zero and variance one, under either of the following conditions:

(6.22) $b = \infty$, (6.17) - (6.20).

(6.23) $b < \infty$, (6.17) - (6.21).

Since the normalizing factor is not $\sqrt{n}$ but $\sqrt{n \log n}$, we are here dealing with a non-regular case, and the classical theory does not apply. We shall prove that $\hat{\theta}_n$ is asymptotically efficient by showing that it is (asymptotically) equivalent to Z_n (with respect to the R defined earlier in this chapter). That the regularity conditions (3.5) and (3.6) are satisfied follows immediately from the fact that $\alpha_n(\hat{\theta}_n - \theta_0)$ has a limiting distribution and θ is a translation parameter.

Thus $\hat{\theta}_n$ is asymptotically efficient for all θ when compared with any competing estimator T_n which satisfies (3.7) with $k(n) = \sqrt{n \log n}$. If, in addition, one requires, as is customary in the literature (but not needed by us), that $\alpha_n(T_n - \theta_0)$ is also asymptotically normally distributed with mean zero and variance $\sigma^2_{\theta_0}(T)$, then we easily have that

$$\sigma^2_{\theta_0}(T) \geqq 1 ,$$

the "classical" result. ("Classical" is in quotation marks because the classical theory does not apply.)

For our problem an m.p. estimator Z_n can be obtained as follows. Let $\{\ell_n\}$ be a sequence of positive numbers which approach zero. Then Z_n satisfies

(6.24)

$$\alpha_n \int_{Z_n - r/\alpha_n}^{Z_n + r/\alpha_n} \prod_{i=1}^{n} f(X_i - \theta)d\theta \geq \alpha_n(1-\ell_n) \sup_d \int_{d-r/\alpha_n}^{d+r/\alpha_n} \prod_{i=1}^{n} f(X_i-\theta)d\theta .$$

Define $t = \alpha_n(\theta - \hat{\theta}_n)$ and

$$(6.25) \qquad V_n(t) = \left[\prod_{i=1}^{n} f(X_i - \theta)\right]\left[\prod_{i=1}^{n} f(X_i - \hat{\theta}_n)\right]^{-1} .$$

Since the second factor of $V_n(t)$ does not depend on θ, we may rewrite (6.24) as

$$(6.26) \qquad \int_{\alpha_n(Z_n-\hat{\theta}_n)-r}^{\alpha_n(Z_n-\hat{\theta}_n)+r} V_n(t)dt \geq (1-\ell_n) \sup_d \int_{\alpha_n(d-\hat{\theta}_n)-r}^{\alpha_n(d-\hat{\theta}_n)+r} V_n(t)dt .$$

It follows from Section 3(9) and the limiting distribution of $\alpha_n(\hat{\theta}_n - \theta_0)$ (obtained in [22]) that the supremum operation in (6.24) may be limited to d in the θ-interval

$$(6.27) \qquad \left(\hat{\theta}_n - \frac{1}{2}\frac{\sqrt[4]{\log n}}{\alpha_n}, \hat{\theta}_n + \frac{1}{2}\frac{\sqrt[4]{\log n}}{\alpha_n}\right) .$$

The t-interval, into which (6.27) is transformed by $t = \alpha_n(\theta - \hat{\theta}_n)$, is contained in the t-interval

(6.28) $\{\hat{\theta}_n - \sqrt[4]{\log n}\ ,\ \hat{\theta}_n + \sqrt[4]{\log n}\ \}$.

It will be shown shortly that Lemmas 3.4 and 3.5 of [22] hold with their constant k replaced by $\sqrt[4]{\log n}$; we assume this for the moment. It then follows from this version of these lemmas that

(6.29) $$\sup_t \frac{\left|\log V_n(t) + \frac{1}{2} t^2\right|}{t^2}$$

converges to 0 in P_{θ_0}-probability as $n \to \infty$; the supremum in (6.29) is with respect to t in the t-interval (6.28). (When $t = 0$ the expression being maximized in (6.29) becomes $\frac{0}{0}$, and we define it as 0.) From the above we conclude that, if we set $Z_n = \hat{\theta}_n$, there exists a sequence $\{\ell_n\}$ such that (6.26) is satisfied with P_{θ_0}-probability approaching one. This proves that $\hat{\theta}_n$ is asymptotically equivalent to the m.p. estimator Z_n.

It remains to prove the extended versions of Lemmas 3.4 and 3.5 of [22]. Define

$$M_n = \min (X_1,\ldots,X_n)$$

$$N_n = \max (X_1,\ldots,X_n) .$$

An examination of the proofs of Lemmas 3.4 and 3.5 of [22] shows that the proofs would remain unchanged if $\sqrt[4]{\log n}$ were substituted for their k, excepting only that we must now show for Lemma 3.4 that

(6.30) $$P_{\theta_0}\left[M_n - \theta_0 \geq \frac{\delta_n}{\varepsilon} = \frac{\log n}{\varepsilon\alpha_n}\right] \to 1 ,$$

and for Lemma 3.5 that, in addition,

$$(6.31) \qquad P_{\theta_0}\left[b - N_n \geq \frac{\delta_n}{1-\beta}\right] \to 1 .$$

For large n the left member of (6.30) is greater than

$$\left[1 - \frac{\alpha\,\delta_n^2}{\varepsilon^2}\right]^n = \left[1 - \frac{2\sqrt{\log n}}{\varepsilon^2 n \log n}\right]^n \to 1 .$$

This proves (6.30). As for (6.31), this statement is weaker than the conclusion of Lemma 2.2 of [22], so it certainly holds.

As mentioned earlier, the $\{\hat{\theta}_n\}$ of [22] is a consistent sequence of roots of the likelihood equation. Let $k_1(\cdot)$ be any function on the positive integers such that $k_1(n) \uparrow \infty$, $n^{-1/2}\,k_1(n) \to 0$. Then any consistent sequence will eventually lie in an interval of length $n^{-1/2}\,k_1(n)$ centered at M_n. This remark may help to identify the root of the likelihood equation which is a member of a consistent sequence.

CHAPTER 7: TESTING HYPOTHESES

In [14], the basic idea used in proving the asymptotic optimality of maximum probability estimators was used to construct a minimax test of a hypothesis about a single parameter in the presence of nuisance parameters, in cases where the observations were independent and identically distributed and certain regularity conditions were satisfied. It was mentioned in [14] that the technique could be applied to testing a hypothesis about several parameters, and to situations where the observations are not necessarily independent and identically distributed. This chapter carries out this extension.

Since the dimension m of Θ is greater than one in this chapter, each component of the vector of parameters requires its own normalizing factor. $k_i(n)$ will be the symbol used for the normalizing factor for the parameter θ_i $(i=1,\ldots,m)$. This differs from the notation used in the earlier chapters, where the discussion was carried out in detail only for the case $m = 1$.

The outline of this chapter is as follows: (1) statement of assumptions and notation; (2) asymptotic distribution theory; (3) solution of an artificial problem where it is known that θ lies in a small neighborhood; (4) solution of the "real" problem; (5) examples.

7(1). Throughout this chapter, we assume the following:

(7.1) There exist m sequences of nonrandom positive quantities, $\{k_1(n)\}, \ldots, \{k_m(n)\}$, with $\lim_{n\to\infty} k_i(n) = \infty$ for $i = 1,\ldots,m$, such that for any $\theta^0 = (\theta_1^0,\ldots,\theta_m^0)$ in Θ,

$$-\frac{1}{k_i(n)k_j(n)}\frac{\partial^2}{\partial\theta_i\partial\theta_j}\log K_n(X(n)|\theta_1,\ldots,\theta_m)]_{\theta^0} \quad \text{converges}$$

stochastically as n increases to a nonrandom quantity, say $B_{ij}(\theta^0)$, when θ^0 is the true parameter value, for $i = 1,\ldots,m$ and $j = 1,\ldots,m$. $B_{ij}(\theta^0)$ is assumed to be a continuous function of θ^0. Let $B(\theta^0)$ denote the m by m matrix with $B_{ij}(\theta^0)$ in row i and column j. We assume $[B(\theta^0)]^{-1}$ exists, and denote it by $I(\theta^0)$.

(7.2) For each θ^0 in Θ, we assume there exist m sequences of nonrandom positive quantities $\{M_1^*(n,\theta^0)\}, \ldots, \{M_m^*(n,\theta^0)\}$, satisfying the following conditions:

(a) $\lim_{n\to\infty} M_i^*(n,\theta^0) = \infty$ for $i = 1,\ldots,m$.

(b) $\lim_{n\to\infty} \dfrac{M_i^*(n,\theta^0)}{k_i(n)} = 0$ for $i = 1,\ldots,m$.

(c) Let $N_n(\theta^0)$ denote the set of all vectors $\theta = (\theta_1,\ldots,\theta_m)$ such that $|\theta_i-\theta_i^0| \leq \dfrac{M_i^*(n,\theta^0)}{k_i(n)}$ for $i = 1,\ldots,m$. For all sufficiently large n, $N_n(\theta^0)$ is contained in Θ. We denote

$$-\frac{1}{k_i(n)k_j(n)} \frac{\partial^2}{\partial\theta_i\partial\theta_j} \log K_n(X(n)|\theta) - B_{ij}(\theta^0) \text{ by } \varepsilon_{ij}(\theta,\theta^0,n).$$

For any $\gamma > 0$, let $S_n(\theta^0,\gamma)$ denote the region in $X(n)$-space where $\sum_{i=1}^{m} \sum_{j=1}^{m} M_i^*(n,\theta^0)M_j^*(n,\theta^0) \sup_{\theta \text{ in } N_n(\theta^0)} |\varepsilon_{ij}(\theta,\theta^0,n)| < \gamma$. We assume that there exist two sequences of nonrandom positive quantities $\{\gamma(n,\theta^0)\}$, $\{\delta(n,\theta^0)\}$, with $\lim_{n\to\infty} \gamma(n,\theta^0) = 0$ and $\lim_{n\to\infty} \delta(n,\theta^0) = 0$, such that for each n and each θ in $N_n(\theta^0)$,

$P_\theta[X(n) \text{ in } S_n(\theta^0,\gamma(n,\theta^0))] > 1 - \delta(n,\theta^0)$.

This completes the list of assumptions. It is shown in the Appendix that in the special case where $m = 1$ and $X(n)$ consists of n independent and identically distributed elements, the assumptions are less restrictive than those given in 6(1).

We now introduce additional notation and terminology that will be used in this chapter.

$\{\Delta_i(n,\theta^0)\}$ is, for each i, some sequence of nonrandom positive quantities, $\Delta_i(n,\theta^0)$ depending only on i, n, and θ^0, with $\lim_{n\to\infty} \Delta_i(n,\theta^0) = 0$. We call such a sequence a "null sequence."

$Q(s)$ will denote the s by s identity matrix.

$\frac{1}{k_i(n)} \frac{\partial}{\partial\theta_i} \log K_n(X(n)|\theta_1,\ldots,\theta_m)]_{\theta^*}$ will be denoted by $A_i(n,\theta^*)$. The row vector $(A_1(n,\theta^*), \ldots, A_m(n,\theta^*))$ will be denoted by $A(n,\theta^*)$.

For typographical simplicity, we use notation illustrated as follows: $B^{-1}(\theta^0)$, $B'(\theta^0)$ denote the inverse and transpose respectively of $B(\theta^0)$.

There exists a nonsingular m by m matrix $C(\theta^0)$, with element in row i and column j denoted by $C_{ij}(\theta^0)$, with $C_{ij}(\theta^0) = 0$ if $j < i$, such that $C'(\theta^0)B(\theta^0)C(\theta^0) = Q(m)$. (See page 343 of [2], for example.)

The one by m vector $\omega(\theta,\theta^0,n) = (\omega_1(\theta,\theta^0,n),\ldots,\omega_m(\theta,\theta^0,n))$ is defined by the equation $\omega(\theta,\theta^0,n)C'(\theta^0) = (k_1(n)(\theta_1-\theta_1^0),\ldots,k_m(n)(\theta_m-\theta_m^0))$. The one by m vector $Z(n,\theta^0) = (Z_1(n,\theta^0),\ldots,Z_m(n,\theta^0))$ is defined by the equation $Z(n,\theta^0) = A(n,\theta^0)C(\theta^0)$.

$G_s(z_1,\ldots,z_s;\mu_1,\ldots,\mu_s)$ denotes the s-variate normal cumulative distribution function with means $\mu_1, \ldots, \mu_s$ and covariance matrix $Q(s)$. $\chi_s(z;p)$ denotes the noncentral chi-square cumulative distribution function with s degrees of freedom and noncentrality parameter ρ.

If $L = (L_1,\ldots,L_s)$ and $U = (U_1,\ldots,U_s)$ are given vectors with $L_i < U_i$ for $i = 1,\ldots,s$, then $S_s(L,U)$ denotes the set of all s-dimensional points $z_1, \ldots, z_s$ such that $L_i < z_i \leq U_i$ for $i = 1,\ldots,s$. L_i may be $-\infty$, U_i may be ∞. $G_s^*(S_s(L,U);\mu_1,\ldots,\mu_s)$ denotes the probability assigned to $S_s(L,U)$ by the distribution function $G_s(z_1,\ldots,z_s;\mu_1,\ldots,\mu_s)$. In general, if F is any cumulative distribution function, and R is any region in the relevant space, $F^*(R)$ denotes the probability assigned to R by F.

If E is any event or set, $\bar{E}$ denotes its complement. The following inequality is called "Bonferroni's inequality": For any events $E_1, \ldots, E_s$, $P(E_1 \cap \cdots \cap E_s) \geq 1 - \sum_{i=1}^{s} P(\bar{E}_i)$.

7(2). For any given θ^0 in Θ and any θ in $N_n(\theta^0)$, a simple Taylor's expansion of $\log K_n(X(n)|\theta)$ around θ^0 and use of the definitions above gives

$$(7.3) \qquad K_n(x(n)|\theta) =$$

$$K_n(x(n)|\theta^0) \exp \left[\begin{array}{l} \omega(\theta,\theta^0,n)Z'(n,\theta^0) \\ - \frac{1}{2}\, \omega(\theta,\theta^0,n)Q(m)\omega'(\theta,\theta^0,n) \\ - \frac{1}{2} \sum_{i=1}^{m} \sum_{j=1}^{m} k_i(n)(\theta_i-\theta_i^0)k_j(n)(\theta_j-\theta_j^0)\varepsilon_{ij}(\vartheta,\theta_0,n) \end{array} \right]$$

where $\vartheta = (\vartheta_1,\ldots,\vartheta_m)$ is in $N_n(\theta^0)$, and $Z(n,\theta^0)$, $\varepsilon_{ij}(\vartheta,\theta^0,n)$ are defined using $x(n)$ in place of $X(n)$.

In (7.3) we integrate (or sum) with respect to the components of $x(n)$ over $S_n(\theta^0,\gamma(n,\theta^0))$, and use assumption (7.2) and the law of the mean for integrals to get that for all θ in $N_r(\theta^0)$,

$$(7.4) \qquad P_\theta[X(n) \text{ in } S_n(\theta^0,\gamma(n,\theta^0))] =$$

$$\exp\{\beta(\theta,\theta^0,n)\gamma(n,\theta^0)\}\int_{S_n(\theta^0,\gamma(n,\theta^0))} K_n(x(n)|\theta^0) \exp\left[\begin{array}{l}\omega(\theta,\theta^0,n)Z'(n,\theta^0) \\ \\ -\frac{1}{2}\omega(\theta,\theta^0,n)\omega'(\theta,\theta^0,n)\end{array}\right] dx(n)$$

where $|\beta(\theta,\theta^0,n)| \leq \frac{1}{2}$. We denote the integral in (7.4) by $\bar{D}(n,\theta,\theta^0)$. Recalling the definition of $S_n(\theta^0,\gamma(n,\theta^0))$, we find $1 - \delta(n,\theta^0) < \exp\{\beta(\theta,\theta^0,n)\gamma(n,\theta^0)\}\bar{D}(n,\theta,\theta^0) \leq 1$, from which it follows that for all θ in $N_n(\theta^0)$, $|\bar{D}(n,\theta,\theta^0) - 1| \leq \Delta_1(n,\theta^0)$.

For each θ in $N_n(\theta^0)$, we define $\bar{K}_n(x(n)|\theta)$ as follows:

$$\bar{K}_n(x(n)|\theta) = \frac{K_n(x(n)|\theta^0)}{\bar{D}(n,\theta,\theta^0)} \exp\left[\begin{array}{l}\omega(\theta,\theta^0,n)Z'(n,\theta^0) \\ \\ -\frac{1}{2}\omega(\theta,\theta^0,n)\omega'(\theta,\theta^0,n)\end{array}\right]$$

$$\text{if } x(n) \text{ is in } S_n(\theta^0,\gamma(n,\theta^0))$$

$$= 0 \text{ otherwise.}$$

Thus $\bar{K}_n(x(n)|\theta)$ is a density function. $\bar{P}_\theta(E)$ will denote the probability of the event E under the assumption that the distribution of $X(n)$ is given by $\bar{K}_n(x(n)|\theta)$. $\bar{E}_\theta(V)$ will denote the expectation of V under the assumption that the distribution of $X(n)$ is given by $\bar{K}_n(x(n)|\theta)$.

For any θ in $N_n(\theta^0)$,

$$\bar{P}_\theta\left[\bar{D}(n,\theta,\theta^0)\, e^{-\frac{1}{2}\gamma(n,\theta^0)} \le \frac{K_n(X(n)|\theta)}{\bar{K}_n(X(n)|\theta)} \le \bar{D}(n,\theta,\theta^0)e^{\frac{1}{2}\gamma(n,\theta^0)}\right] = 1,$$

and it follows that if $R(n)$ is any measurable region in $X(n)$-space,

$$(7.5)\qquad |\bar{P}_\theta(X(n) \text{ in } R(n)) - P_\theta(X(n) \text{ in } R(n))| \le$$

$$2 \max \begin{bmatrix} \left|1 - \bar{D}(n,\theta,\theta^0)e^{-\frac{\gamma(n,\theta^0)}{2}}\right| \\ \\ \left|1 - \bar{D}(n,\theta,\theta^0)e^{\frac{\gamma(n,\theta^0)}{2}}\right| \end{bmatrix} \le \Delta_2(n,\theta^0), \quad \text{say},$$

where $\lim_{n\to\infty} \Delta_2(n,\theta^0) = 0$.

Let $H_n(z_1,\ldots,z_m;\theta,\theta^0)$ denote the joint cumulative distribution function for the components of $Z(n,\theta^0)$ when $X(n)$ has distribution $K_n(x(n)|\theta)$. Let $\bar{H}_n(z_1,\ldots,z_m;\theta,\theta^0)$ denote the joint cumulative distribution function for the components of $Z(n,\theta^0)$ when $X(n)$ has distribution $\bar{K}_n(x(n)|\theta)$. Let $h_n(z_1,\ldots,z_m;\theta,\theta^0)$, $\bar{h}_n(z_1,\ldots,z_m;\theta,\theta^0)$ denote the respective density functions corresponding to H_n, $\bar{H}_n$.

For each θ in $N_n(\theta^0)$, we have

$$\bar{K}_n(X(n)|\theta) = \frac{\bar{D}(n,\theta^0,\theta^0)}{\bar{D}(n,\theta,\theta^0)}\, \bar{K}_n(X(n)|\theta^0) \exp \begin{bmatrix} \omega(\theta,\theta^0,n)Z'(n,\theta^0) \\ \\ -\frac{1}{2}\omega(\theta,\theta^0,n)\omega'(\theta,\theta^0,n) \end{bmatrix}$$

from which it follows that for all $z = (z_1,\ldots,z_m)$, and all θ

in $N_n(\theta^0)$,

(7.6)

$$\overline{h}_n(z;\theta,\theta^0) = \frac{\overline{D}(n,\theta^0,\theta^0)}{\overline{D}(n,\theta,\theta^0)} \overline{h}_n(z;\theta^0,\theta^0) \exp\left[\omega(\theta,\theta^0,n)z' - \frac{1}{2}\omega(\theta,\theta^0,n)\omega'(\theta,\theta^0,n)\right].$$

Lemma 7.1.

$$\lim_{n\to\infty} \sup_{L,U} \left|\overline{H}^*_n(S_m(L,U);\theta^0,\theta^0) - G^*_m(S_m(L,U);0,\ldots,0)\right| = 0 .$$

Proof. Choose arbitrary finite values $(t_1,\ldots,t_m)$ and hold them fixed. Define $\theta(n) = (\theta_1(n),\ldots,\theta_m(n))$ by the equation $\omega(\theta(n),\theta^0,n) = (t_1,\ldots,t_m)$. For all sufficiently large n, $\theta(n)$ will be in $N_n(\theta^0)$. Replace θ in (7.6) by $\theta(n)$, and integrate both sides of (7.6) with respect to $(z_1,\ldots,z_m)$, getting

$$1 = \frac{\overline{D}(n,\theta^0,\theta^0)}{\overline{D}(n,\theta(n),\theta^0)} \exp\left[-\frac{1}{2}\sum_{i=1}^{m}\sum_{j=1}^{m} t_i t_j\right] \overline{E}_{\theta^0}\left[\exp\left\{\sum_{i=1}^{m} t_i Z_i(n,\theta^0)\right\}\right]$$

which implies

$$\lim_{n\to\infty} \overline{E}_{\theta^0}(\exp\{\sum_{i=1}^{m} t_i Z_i(n,\theta^0)\}) = \exp(\frac{1}{2}\sum_{i=1}^{m}\sum_{j=1}^{m} t_i t_j) ,$$

which in turn implies

$$\lim_{n\to\infty} \sup_{z_1,\ldots,z_m} \{|\overline{H}_n(z_1,\ldots,z_m;\theta^0,\theta^0) - G_m(z_1,\ldots,z_m;0, \ldots,0)|\} = 0 .$$

The proof of the lemma is an immediate caonsequence of this last equality.

Theorem 7.2. For each θ^0 in Θ, there exist m sequences of nonrandom positive quantities $\{\overline{M}_1(n,\theta^0)\}, \ldots, \{\overline{M}_m(n,\theta^0)\}$, with $\overline{M}_i(n,\theta^0) \leq M_i^*(n,\theta^0)$ and $\lim_{n\to\infty} \overline{M}_i(n,\theta^0) = \infty$ for $i = 1,\ldots,m$, such that

$$\lim_{n\to\infty} \sup_{\theta \text{ in } \overline{N}_n(\theta^0)} \{\sup_{L,U} |\overline{H}_n^*(S_m(1,U);\theta,\theta^0) - G_m^*(S_m(L,U);\omega(\theta,\theta^0,n))|\} = 0,$$

where $\overline{N}_n(\theta^0)$ is the set of all vectors $\theta = (\theta_1,\ldots,\theta_m)$ such that $|\theta_i - \theta_i^0| \leq \dfrac{\overline{M}_i(n,\theta^0)}{k_i(n)}$ for $i = 1,\ldots,m$.

Proof. From Lemma 7.1,

$$\sup_{L,U} \{|\overline{H}_n^*(S_m(L,U);\theta^0,\theta^0) - G_m^*(S_m(1,U);0,\ldots,0)|\} \leq \Delta_3(n,\theta^0), \quad \text{say.}$$

From (7.6), we get, for any θ in $N_n(\theta^0)$,

(7.7) $$\overline{H}_n^*(S_m(L,U);\theta,\theta^0) =$$

$$\frac{\overline{D}(n,\theta^0,\theta^0)}{\overline{D}(n,\theta,\theta^0)} \exp\{-\tfrac{1}{2}\omega(\theta,\theta^0,n)\omega'(\theta,\theta^0,n)\} \times$$

$$\times \int_{L_m}^{U_m} \cdots \int_{L_1}^{U_1} \exp\{\omega(\theta,\theta^0,n)z'\} d_{z_1,\ldots,z_m} \overline{H}_n(z;\theta^0,\theta^0) .$$

We evaluate the integral in (7.7) by m-dimensional integration by parts: the formula for $m = 1$ is familiar; the formula for $m = 2$ (using simplified notation) is

$$\int_{L_2}^{U_2}\int_{L_1}^{U_1} r(z_1,z_2)d_{z_1,z_2}\bar{H}(z_1,z_2) =$$

$$\int_{L_2}^{U_2}\int_{L_1}^{U_1} \bar{H}(z_1,z_2)d_{z_1,z_2}r(z_1,z_2) + \int_{L_2}^{U_2} \bar{H}(L_1,z_2)d_{z_2}r(L_1,z_2)$$

$$+ \int_{L_1}^{U_1} \bar{H}(z_1,L_2)d_{z_1}r(z_1,L_2) - \int_{U_2}^{L_2} \bar{H}(U_1,z_2)d_{z_2}r(U_1,z_2)$$

$$- \int_{L_1}^{U_1} \bar{H}(z_1,U_2)d_{z_1}r(z_1,U_2) + \bar{H}(L_1,L_2)r(L_1,L_2)$$

$$+ \bar{H}(U_1,U_2)r(U_1,U_2) - \bar{H}(L_1,U_2)r(L_1,U_2) - \bar{H}(U_1,L_2)r(U_1,L_2)$$

with analogous formulas for $m > 2$. For any m, the result is to eliminate the operation of taking differentials of $\bar{H}_n(z;\theta^0,\theta^0)$. Define $V_n(z;\theta^0)$ as $\bar{H}_n(z;\theta^0,\theta^0) - G_m(z;0,\ldots,0)$. After integrating by parts in (7.7), we replace $\bar{H}_n(z;\theta^0,\theta^0)$ by $G_m(z;0,\ldots,0) + V_n(z;\theta^0)$. The resulting terms involving $G_m(z;0,\ldots,0)$ give the original integral with $\bar{H}_n(z;\theta^0,\theta^0)$ replaced by $G_m(z;0,\ldots,0)$. The terms involving $V_n(z;\theta^0)$ can be written as $\Delta_3(n,\theta^0)T_n(\omega(\theta,\theta^0,n),L,U,\theta^0)$ where $|T_n(\omega(\theta,\theta^0,n),L,U,\theta^0)|$ remains bounded if all components of $\omega(\theta,\theta^0,n)$, L, and U are bounded in absolute value. Thus (7.7) becomes

(7.8) $$\bar{H}_n^*(S_m(L,U);\theta,\theta^0) =$$

$$\frac{\bar{D}(n,\theta^0,\theta^0)}{\bar{D}(n,\theta,\theta^0)} \exp\{-\tfrac{1}{2}\omega(\theta,\theta^0,n)\omega'(\theta,\theta^0,n)\} \times$$

$$\times \left[\int_{L_m}^{U_m} \cdots \int_{L_1}^{U_1} \exp\{\omega(\theta,\theta^0,n)z'\} d_{z_1,\ldots,z_m} G_m(z;0,\ldots,0) + \Delta_3(n,\theta^0) T_n(\omega(\theta,\theta^0,n),L,U,\theta^0) \right]$$

Since $d_{z_1,\ldots,z_m} G_m(z;0,\ldots,0) = \frac{1}{(2\pi)^{m/2}} \exp\{-\frac{1}{2} zz'\}$, it is easily seen that

$$\exp\{-\tfrac{1}{2}\omega(\theta,\theta^0,n)\omega'(\theta,\theta^0,n)\}\exp\{\omega(\theta,\theta^0,n)z'\}d_{z_1,\ldots,z_m}G_m(z;0,\ldots,0)$$

$= d_{z_1,\ldots,z_m} G_m(z;\omega(\theta,\theta^0,n))$, and using this result in (7.8), we get, for every θ in $N_n(\theta^0)$,

(7.9) $$\bar{H}_n^*(S_m(L,U);\theta,\theta^0) = \frac{\bar{D}(n,\theta^0,\theta^0)}{\bar{D}(n,\theta,\theta^0)} G_m^*(S_m(L,U);\omega(\theta,\theta^0,n))$$

$$+ \Delta_3(n,\theta^0)T_n(\omega(\theta,\theta^0,n),L,U,\theta^0)\frac{\bar{D}(n,\theta^0,\theta^0)}{\bar{D}(n,\theta,\theta^0)}\exp\{-\tfrac{1}{2}\omega(\theta,\theta^0,n)\omega'(\theta,\theta^0,n)\} .$$

Recalling the properties of $T_n(\omega(\theta,\theta^0,n),L,U,\theta^0)$ and $\bar{D}(n,\theta,\theta^0)$, it follows from (7.9) that we can find $3m$ sequences of nonrandom quantities, $\{L_1(n,\theta^0)\}$, ..., $\{L_m(n,\theta^0)\}$, $\{U_1(n,\theta^0)\}$, ..., $\{U_m(n,\theta^0)\}$, $\{\bar{M}_1(n,\theta^0)\}$, ..., $\{\bar{M}_m(n,\theta^0)\}$, with $L_i(n,\theta^0) < U_i(n,\theta^0)$, $\lim_{n\to\infty} L_i(n,\theta^0) = -\infty$, $\lim_{n\to\infty} U_i(n,\theta^0) = \infty$, $\bar{M}_i(n,\theta^0) \leq M_i^*(n,\theta^0)$,

$\lim_{n\to\infty} \overline{M}_i(n,\theta^0) = \infty$ (all for $i = 1,\ldots,m$) such that for all n, all θ in $\overline{N}_n(\theta^0)$, and all L, U with $L_i(n,\theta^0) \leq L_i < U_i \leq U_i(n,\theta^0)$ $(i = 1,\ldots,m)$, $|\overline{H}_n^*(S_m(L,U);\theta,\theta^0) - G_m^*(S_m(L,U);\omega(\theta,\theta^0,n))| \leq \Delta_4(n,\theta^0)$, and $G_m^*(S_m(L(n,\theta^0),U(n,\theta^0));\omega(\theta,\theta^0,n)) \geq 1 - \Delta_5(n,\theta^0)$, for some null sequences $\{\Delta_4(n,\theta^0)\}$, $\{\Delta_5(n,\theta^0)\}$.

Now let $L = (L_1,\ldots,L_m)$, $U = (U_1,\ldots,U_m)$ be arbitrary, except that $L_i < U_i$ for $i = 1,\ldots,m$. Define $S(1)$ as $S_m(L,U) \cap S_m(L(n,\theta^0),U(n,\theta^0))$, and $S(2)$ as $S_m(L,U) \cap \overline{S_m(L(n,\theta^0),U(n,\theta^0))}$. Thus $S_m(L,U) = S(1) \cup S(2)$, where $S(1)$, $S(2)$ are disjoint. $S(1)$ is either empty, or else is equal to $S_m(\overline{L}, \overline{U})$, where $L_i(n,\theta^0) \leq \overline{L}_i < \overline{U}_i \leq U_i(n,\theta^0)$. Then, for all θ in $\overline{N}_n(\theta^0)$, we have the following inequalities:

$$G_m^*(S(2);\omega(\theta,\theta^0,n)) \leq \Delta_5(n,\theta^0);$$

$$\overline{H}_n^*(S(2);\theta,\theta^0) \leq \Delta_4(n,\theta^0) + \Delta_5(n,\theta^0);$$

$$|\overline{H}_n^*(S(1);\theta,\theta^0) - G_m^*(S(1);\omega(\theta,\theta^0,n))| \leq \Delta_4(n,\theta^0).$$

It follows that $|\overline{H}_n^*(S_m(L,U);\theta,\theta^0) - G_m^*(S_m(L,U);\omega(\theta,\theta^0,n))| \leq 2[\Delta_4(n,\theta^0) + \Delta_5(n,\theta^0)]$ for all θ in $\overline{N}_n(\theta^0)$. This proves Theorem 7.2.

<u>Theorem 7.3</u>. For any given integer s $(0 \leq s \leq m-1)$ and any given positive value x,

$$\lim_{n\to\infty} \sup_{\theta \text{ in } \overline{N}_n(\theta^0)} |\overline{P}_\theta(\sum_{i=s+1}^{m} Z_i^2(n,\theta^0) \leq x) - \chi_{m-s}(x;[\sum_{i=s+1}^{m} \omega_i^2(\theta,\theta^0,n)]^{\frac{1}{2}})| = 0 .$$

Proof. Let $\bar{J}_n(y_1,\ldots,y_{m-s};\theta,\theta^0)$ denote the joint cumulative distribution function for $Z_{s+1}(n,\theta^0)$, ..., $Z_m(n,\theta^0)$, when $X(n)$ has distribution $\bar{K}_n(x(n)|\theta)$. It is a direct consequence of Theorem 7.2 that

$$(7.10)\qquad \sup_{\theta \text{ in } \bar{N}_n(\theta^0)} \sup_{L,U} |\bar{J}^*_n(S_{m-s}(L,U);\theta,\theta^0) - G^*_{m-s}(S_{m-s}(L,U);\omega_{s+1}(\theta,\theta^0,n),\ldots,\omega_m(\theta,\theta^0,n))| \leq \Delta_6(n,\theta^0)$$

for some null sequence $\{\Delta_6(n,\theta^0)\}$.

Denote the set of points $(y_1,\ldots,y_{m-s})$ such that $\sum_{i=1}^{m-s} y_i^2 \leq x$ by $C(x)$. For any given $\varepsilon > 0$, there are sets $T_1(\varepsilon)$, $T_2(\varepsilon)$ in (m-s)-dimensional space with the following properties:

(7.11)

(a) $T_1(\varepsilon) \subset C(x) \subset T_2(\varepsilon)$.

(b) $G^*_{m-s}(T_2(\varepsilon);\mu_1,\ldots,\mu_{m-s}) - G^*_{m-s}(T_1(\varepsilon);\mu_1,\ldots,\mu_{m-s}) \leq \varepsilon$ for all $\mu_1,\ldots,\mu_{m-s}$.

(c) $T_1(\varepsilon)$, $T_2(\varepsilon)$ is each the union of no more than $q(\varepsilon)$ disjoint sets of the type $S_{m-s}(L,U)$, where $q(\varepsilon) < \infty$.

Using (7.10) and (7.11c), we have

(7.12)

$$\sup_{\theta \text{ in } \bar{N}_n(\theta^0)} |\bar{J}_n^*(T_1(\varepsilon);\theta,\theta^0) - G_{m-s}^*(T_1(\varepsilon);\omega_{s+1}(\theta,\theta^0,n),\ldots,\omega_m(\varepsilon,\theta^0,n))|$$

$$\leq q(\varepsilon)\Delta_6(n,\theta^0) \ .$$

Using (7.11a), (7.11b), and (7.12), we get

(7.13)

$$\sup_{\theta \text{ in } \bar{N}_n(\theta^0)} |\bar{J}_n^*(C(x);\theta,\theta^0) - G_{m-s}^*(C(x);\omega_{s+1}(\theta,\theta^0,n),\ldots,\omega_m(\theta,\theta^0,n))|$$

$$\leq \varepsilon + q(\varepsilon)\Delta_6(n,\theta^0) \ .$$

Since $G_{m-s}^*(C(x);\omega_{s+1}(\theta,\theta^0,n),\ldots,\omega_m(\theta,\theta^0,n)) =$ $\chi_{m-s}(x;\ [\sum_{i=s+1}^{m} \omega_i^2(\theta,\theta^0,n)]^{\frac{1}{2}})$, Theorem 7.3 follows from (7.13).

Theorem 7.4. Theorem 7.2 holds with $\bar{H}_n^*(S_m(L,U);\theta,\theta^0)$ replaced by $H_n^*(S_m(L,U);\theta,\theta^0)$; and Theorem 7.3 holds with $\bar{P}_\theta(\sum_{i=s+1}^{m} Z_i^2(n,\theta^0) \leq x)$ replaced by $P_\theta(\sum_{i=s+1}^{m} Z_i^2(n,\theta^0) \leq x)$.

Proof. The theorem follows directly from (7.5) and the fact that $\bar{N}_n(\theta^0)$ is contained in $N_n(\theta^0)$.

7(3). For each positive integer n, we introduce the following statistical problem. We observe $X(n)$, which has distribution given by $K_n(x(n)|\theta(n))$, where $\theta(n) = (\theta_1(n),\ldots,\theta_m(n))$ is unknown, and the problem is to test the hypothesis that

$\theta_{s+1}(n) = \theta^0_{s+1}, \ldots, \theta_m(n) = \theta^0_m$, where $\theta^0_{s+1}, \ldots, \theta^0_m$ are given known values.

Throughout 7(3), we are going to consider an "artificial" version of the problem just described, a version made artificial by assuming that it is known that $\theta(n)$ is in $\overline{N}_n(\theta^0)$ for each n, where $\theta^0_1, \ldots, \theta^0_s$ are given known values such that $\theta^0 = (\theta^0_1,\ldots,\theta^0_m)$ is in Θ.

Until further notice, we are going to assume that the joint distribution of $X(n)$ is given by $\overline{K}_n(x(n)|\theta)$, and we replace the parameters $(\theta_1(n),\ldots,\theta_m(n))$ by the set of parameters $\omega(n) = (\omega_1(n),\ldots,\omega_m(n))$ defined by the equation $(\omega_1(n),\ldots,\omega_m(n))C'(\theta^0) = (k_1(n)(\theta_1(n)-\theta^0_1),\ldots,k_m(n)(\theta_m(n)-\theta^0_m))$. Let $\tilde{N}_n(\theta^0)$ denote the set of all points given by $\omega(n)$ as $\theta(n)$ varies over $\overline{N}_n(\theta^0)$. Any given point $\omega = (\omega_1,\ldots,\omega_m)$ will be in $\tilde{N}_n(\theta^0)$ for all sufficiently large n. Because of the triangular form of $C(\theta^0)$, it is easily verified that the hypothesis that $\theta_{s+1}(n) = \theta^0_{s+1},\ldots,\theta_m(n) = \theta^0_m$ is equivalent to the hypothesis that $\omega_{s+1}(n) = 0,\ldots,\omega_m(n) = 0$.

Now we construct a Bayes test for the hypothesis $\omega_{s+1}(n) = 0,\ldots,\omega_m(n) = 0$, relative to the following a priori distribution for $\omega(n)$. The a priori distribution assigns total probability b to the point $\omega_1(n) = \overline{\omega}_1,\ldots,\omega_s(n) = \overline{\omega}_s,\omega_{s+1}(n) = 0,\ldots,\omega_m(n) = 0$, where $\overline{\omega}_1,\ldots,\overline{\omega}_s$ are arbitrary values, and $0 < b < 1$. The a priori distribution assigns total probability $(1-b)$ to the set of points $S(\delta) \equiv \{(\overline{\omega}_1,\ldots,\overline{\omega}_s,\omega_{s+1},\ldots,\omega_m) : \sum_{i=s+1}^{m} \omega_i^2 = \delta\}$, where $\delta > 0$ is arbitrary. The probability $(1-b)$ is distributed over $S(\delta)$ as follows. If ω is in $S(\delta)$, and $\theta(n)$ denotes the point satisfying $\omega C'(\theta^0) = (k_1(n)(\theta_1(n)-\theta^0_1),\ldots,k_m(n)(\theta_m(n)-\theta^0_m))$, then the a priori density at ω is $\rho_n \bar{D}(n, \theta(n), \theta^0)$, where ρ_n is an appropriate value. ρ_n depends on n, $\bar{\omega}_1, \ldots, \bar{\omega}_s$, θ^0, and δ, but not on $\theta(n)$. We note that for all sufficiently large n, this a priori distribution assigns all probability to $\tilde{N}_n(\theta^0)$.

Recalling that we are assuming that the joint distribution of $X(n)$ is given by $\bar{K}_n(x(n)|\theta)$, a Bayes test relative to the given a priori distribution is as follows, where $\bar{\theta}(n)$ is given by the equation

$$(\bar{\omega}_1, \ldots, \bar{\omega}_s, 0, \ldots, 0)C'(\theta^0) = (k_1(n)(\bar{\theta}_1(n) - \theta_1^0), \ldots, k_m(n)(\bar{\theta}_m(n) - \theta_m^0)):$$

(7.14) Accept the hypothesis if and only if

$$\int_{\sum_{i=s+1}^{m} \omega_i^2 = \delta} \cdots \int \exp\{\sum_{i=s+1}^{m} \omega_i Z_i(n;\theta^0)\} d\omega_{s+1} \cdots d\omega_m \leqq \frac{b e^{\delta/2}}{\rho_n (1-b) \bar{D}(n, \bar{\theta}(n), \theta^0)} .$$

As n increases, $\bar{D}(n, \bar{\theta}(n), \theta^0)$ approaches one, and ρ_n approaches the reciprocal of the surface area of an $(m-s)$-dimensional sphere with radius $\delta^{1/2}$. Let $\sigma(\delta)$ denote this limiting value of ρ_n. The integral in (7.14) is known to be equal to

$\delta^{\frac{m-s}{2}} \Phi(\delta \sum_{i=s+1}^{m} Z_i^2(n, \theta^0))$, where $\Phi(y)$ is positive, continuous, and strictly increasing in y for all $y \geqq 0$. For any given α $(0 < \alpha < 1)$, define $T(\alpha)$ by the equation $\chi_{m-s}(T(\alpha); 0) = 1 - \alpha$. Define $b(\alpha, \delta)$ as the solution in b of the equation

$\frac{b e^{\delta/2} \delta^{(s-m)/2}}{\sigma(\delta)(1-b)} = \Phi(\delta T(\alpha))$. Clearly, $0 < b(\alpha, \delta) < 1$. Now we replace b by $b(\alpha, \delta)$ in (7.14), and call the resulting test T_n^*.

It follows from Theorém 7.3 that

$$\lim_{n\to\infty} \overline{P}_{\overline{\theta}(n)} [T_n^* \text{ rejects the hypothesis}] = \alpha$$

(7.15)

$$\lim_{n\to\infty} \sup_{\theta \text{ in } S_n^*(\delta)} |\overline{P}_\theta(T_n^* \text{ rejects the hypothesis}) - [1-\chi_{m-s}(T(\alpha);\delta^{1/2})]| = 0$$

where $S_n^*(\delta)$ is the set of vectors $\theta = (\theta_1,\ldots,\theta_m)$ given by the relation $\omega C'(\theta^0) = (k_1(n)(\theta_1-\theta_1^0), \ldots, k_m(n)(\theta_m-\theta_m^0))$ as ω varies over $S(\delta)$.

Now let $\overline{T}_n$ be any sequence of tests. It follows from (7.15) and the fact that T_n^* is a Bayes decision rule that if $\lim_{n\to\infty} \overline{P}_{\overline{\theta}(n)} [\overline{T}_n \text{ rejects the hypothesis}] \leq \alpha$, then

$$\overline{\lim_{n\to\infty}} \sup_{\theta \text{ in } S_n^*(\delta)} \overline{P}_\theta(\overline{T}_n \text{ rejects the hypothesis}) \leq 1-\chi_{m-s}(T(\alpha);\delta^{1/2}).$$

The analysis above has been carried out assuming that the distribution of $X(n)$ is given by $\overline{K}_n(x(n)|\theta)$. Now we show that the same results hold when the distribution of $X(m)$ is given by $K_n(x(n)|\theta)$. That (7.15) holds with $\overline{P}_\theta$ replaced by P_θ follows directly from (7.5). Now suppose that there is a sequence $\{\overline{T}_n\}$ of tests such that $\lim_{n\to\infty} P_{\overline{\theta}(n)} [\overline{T}_n \text{ rejects the hypothesis}] \leq \alpha$, and $\overline{\lim_{n\to\infty}} \sup_{\theta \text{ in } S_n^*(\delta)} P_\theta(\overline{T}_n \text{ rejects the hypothesis}) >$ $1-\chi_{m-s}(T(\alpha);\delta^{1/2})$. Then by (7.5) the same would be true if P_θ were replaced by $\overline{P}_\theta$, which would contradict the result of the preceding paragraph.

In the analysis above, the noncentrality parameter in the

expression for the asymptotic power of T_n^* is $\delta^{1/2} = [\sum_{i=s+1}^{m} \omega_i^2]^{1/2}$. Now we express this in terms of θ. To do this, let $D(\theta^0)$ denote the inverse of $C(\theta^0)$. Then $D(\theta^0)$ is triangular in the same way as $C(\theta^0)$ is triangular: if $D_{ij}(\theta^0)$ is the element in row i and column j of $D(\theta^0)$, $D_{ij}(\theta^0) = 0$ if $j < i$.

Next we partition $C(\theta^0)$, $D(\theta^0)$, $B(\theta^0)$, $I(\theta^0)$ as follows:

$$C(\theta^0) = \begin{bmatrix} C(1,1;\theta^0) & C(1,2;\theta^0) \\ C(2,1;\theta^0) & C(2,2;\theta^0) \end{bmatrix}$$

$$D(\theta^0) = \begin{bmatrix} D(1,1;\theta^0) & D(1,2;\theta^0) \\ D(2,1;\theta^0) & D(2,2;\theta^0) \end{bmatrix}$$

$$B(\theta^0) = \begin{bmatrix} B(1,1;\theta^0) & B(1,2;\theta^0) \\ B(2,1;\theta^0) & B(2,2;\theta^0) \end{bmatrix}$$

$$I(\theta^0) = \begin{bmatrix} I(1,1;\theta^0) & I(1,2;\theta^0) \\ I(2,1;\theta^0) & I(2,2;\theta^0) \end{bmatrix}$$

where all northwest corners are s by s, northeast corners are s by $(m-s)$, southwest corners are s by $(m-s)$, and southeast corners are $(m-s)$ by $(m-s)$. $C(2,1;\theta^0)$ and $D(2,1;\theta^0)$ consist of zeroes.

Writing $\begin{bmatrix} D(1,1;\theta^0) & D(1,2;\theta^0) \\ 0 & D(2,2;\theta^0) \end{bmatrix} \begin{bmatrix} C(1,1;\theta^0) & C(1,2;\theta^0) \\ 0 & C(2,2;\theta^0) \end{bmatrix} =$

$\begin{bmatrix} Q(s) & 0 \\ 0 & Q(m-s) \end{bmatrix}$, and equating southeast corners after multiplying, we find $C^{-1}(2,2;\theta^0) = D(2,2;\theta^0)$.

Since $C'(\theta^0)B(\theta^0)C(\theta^0) = Q(m)$, we have $B(\theta^0) = D'(\theta^0)D(\theta^0)$, and $I(\theta^0) = C(\theta^0)C'(\theta^0)$. Writing

$$\begin{bmatrix} I(1,1;\theta^0) & I(1,2;\theta^0) \\ I(2,1;\theta^0) & I(2,2;\theta^0) \end{bmatrix} =$$

$$\begin{bmatrix} C(1,1;\theta^0) & C(1,2;\theta^0) \\ 0 & C(2,2;\theta^0) \end{bmatrix} \begin{bmatrix} C'(1,1;\theta^0) & 0 \\ C'(1,2;\theta^0) & C'(2,2;\theta^0) \end{bmatrix}$$

and equating southeast corners after multiplying, we find $I(2,2;\theta^0) = C(2,2;\theta^0)\ C'(2,2;\theta^0)$ so that $I^{-1}(2,2;\theta^0) = D'(2,2;\theta^0)\ D(2,2;\theta^0)$. We denote the element in row i and column j of $I^{-1}(2,2;\theta^0)$ by $v_{ij}(\theta^0)$, for $i,j = 1,\ldots,m-s$.

Now we write the vector $\omega = (\omega_1,\ldots,\omega_m)$ as $(\omega(1),\omega(2))$, where $\omega(1)$ is one by s. Then the noncentrality parameter $[\sum_{i=s+1}^{m} \omega_i^2]^{1/2}$ can be written as $[\omega(2)\omega'(2)]^{1/2}$. From above, we have $(\omega(1),\omega(2)) =$

$$(k_1(n)(\theta_1-\theta_1^0),\ldots,k_m(n)(\theta_m-\theta_m^0)) \begin{bmatrix} D'(1,1;\theta^0) & 0 \\ D'(1,2;\theta^0) & D'(2,2;\theta^0) \end{bmatrix}$$ from

which we find

$\omega(2) = (k_{s+1}(n)(\theta_{s+1}-\theta^0_{s+1}),\ldots,k_m(n)(\theta_m-\theta^0_m))D'(2,2;\theta^0)$. If follows that $[\sum_{i=s+1}^{m} \omega_i^2]^{1/2} =$

$[\sum_{i=1}^{m-s} \sum_{j=1}^{m-s} v_{ij}(\theta^0)k_{s+i}(n)(\theta_{s+i}-\theta^0_{s+i})k_{s+j}(n)(\theta_{s+j}-\theta^0_{s+j})]^{1/2}$. This is the noncentrality parameter in terms of θ.

7(4). Now we return to the nonartificial problem. For each n, we observe $X(n)$, which has distribution given by $K_n(x(n)|\theta(n))$, where $\theta(n) = (\theta_1(n),\ldots,\theta_m(n))$ is unknown, and we want to test the hypothesis that $\theta_{s+1}(n) = \theta^0_{s+1},\ldots,\theta_m(n) = \theta^0_m$, with asymptotic level of significance α. $\theta^0_{s+1}, \ldots, \theta^0_m$ are given known values, but we know nothing about $\theta_1(n), \ldots, \theta_s(n)$.

Throughout 7(4), we assume that for each n we have available estimators $\tilde{\theta}_1(n), \ldots, \tilde{\theta}_s(n)$ of $\theta_1(n), \ldots, \theta_s(n)$ respectively, with the property that for any given $\theta^* = (\theta^*_1,\ldots,\theta^*_m)$ in the interior of Θ,

$$(7.16)\qquad \sup_{\theta \text{ in } N_n(\theta^*)} P_\theta[k_i(n)|\tilde{\theta}_i(n)-\theta_i| \leq \tfrac{1}{2}\min_j\{M^*_j(n,\theta^*)\};\ i=1,\ldots,s] \geq 1 - \Delta_7(n,\theta^*)$$

for some null sequence $\{\Delta_7(n,\theta^*)\}$. $\tilde{\theta}_1(n), \ldots, \tilde{\theta}_s(n)$ do not necessarily have to be functions of $X(n)$.

Throughout 7(4) we also assume that $\theta(n)$ is given as follows:

(7.17) $(\bar{\omega}_1,\ldots,\bar{\omega}_s,\omega_{s+1},\ldots,\omega_m)C'(\theta^0) =$

$$(k_1(n)(\theta_1(n)-\theta_1^0),\ldots,k_m(n)(\theta_m(n)-\theta_m^0)) ,$$

where $\bar{\omega}_1, \ldots, \bar{\omega}_s$ are arbitrary fixed values, and $\omega_{s+1}, \ldots, \omega_m$ are arbitrary fixed values with $\sum_{i=s+1}^{m} \omega_i^2 = \delta$. θ^0 is in the interior of Θ.

Denote the vector $(\tilde{\theta}_1(n),\ldots,\tilde{\theta}_s(n),\theta^0_{s+1},\ldots,\theta^0_m)$ by $\tilde{\theta}(n)$. $\tilde{\theta}(n)$ is known for each n. Let $\tilde{B}(n)$ denote the m by m matrix with element in row i and column j given by $-\frac{1}{k_i(n)k_j(n)} \frac{\partial^2}{\partial\theta_i\partial\theta_j} \log K_n(X(n)|\theta)]_{\tilde{\theta}(n)}$. It follows from (7.16) and (7.17) that $P_{\theta(n)}[\tilde{B}(n)$ is nonsingular$] > 1 - \Delta_8(n,\theta^0)$ for some null sequence $\{\Delta_8(n,\theta^0)\}$. Let $\tilde{I}(n)$ denote $[\tilde{B}(n)]^{-1}$, and let $\tilde{C}(n)$ denote the m by m triangular matrix with element in row i and column j equal to zero if $j < i$, such that $\tilde{C}'(n)\tilde{B}(n)\tilde{C}(n) = Q(m)$. Define the one by m vector $\tilde{Z}(n) = (\tilde{Z}_1(n),\ldots,\tilde{Z}_m(n))$ by the equation $\tilde{Z}(n) = A(n,\tilde{\theta}(n))\tilde{C}(n)$.

Let $W(n)$ denote the following test procedure. Reject the hypothesis if and only if $\sum_{i=s+1}^{m} \tilde{Z}_i^2(n) > T(\alpha)$. The rest of 7(4) is devoted to investigating the asymptotic properties of $W(n)$.

We write $Z(n,\theta)$ as $(Z(1;n,\theta), Z(2;n,\theta))$, $A(n,\theta)$ as $(A(1;n,\theta), A(2;n,\theta))$, and $\tilde{Z}(n)$ as $(\tilde{Z}(1;n), \tilde{Z}(2;n))$, where first elements are 1 by s and second elements are 1 by $(m-s)$. From the definition of $Z(n,\theta)$, we have

$$(Z(1;n,\theta), Z(2;n,\theta)) = (A(1;n,\theta), A(2;n,\theta)) \begin{bmatrix} C(1,1;\theta) & C(1,2;\theta) \\ 0 & C(2,2;\theta) \end{bmatrix}$$

from which we get $Z'(2;n,\theta) = A(1;n,\theta)C(1,2;\theta) + A(2;n,\theta)C(2,2;\theta)$, and so

$$(7.18)\qquad \sum_{i=s+1}^{m} z_i^2(n,\theta^0) = Z(2;n,\theta^0)Z'(2;n,\theta^0) =$$

$$A(1;n,\theta^0)C(1,2;\theta^0)C'(1,2;\theta^0)A'(1;n,\theta^0)$$

$$+ A(2;n,\theta^0)C(2,2;\theta^0)C'(2,2;\theta^0)A'(2;n,\theta^0)$$

$$+ 2A(1;n,\theta^0)C(1,2;\theta^0)C'(2,2;\theta^0)A'(2;n,\theta^0) .$$

Similarly, we partition $\tilde{C}(n)$ as $\begin{bmatrix} \tilde{C}(1,1;n) & \tilde{C}(1,2;n) \\ 0 & \tilde{C}(2,2;n) \end{bmatrix}$, and we can write

$$(7.19)\qquad \sum_{i=s+1}^{m} \tilde{z}_i^2(n) = A(1;n,\tilde{\theta}(n))\tilde{C}(1,2;n)\tilde{C}'(1,2;n)A'(1;n,\tilde{\theta}(n))$$

$$+ A(2;n,\tilde{\theta}(n))\tilde{C}(2,2;n)\tilde{C}'(2,2;n)A'(2;n,\tilde{\theta}(n))$$

$$+ 2A(1;n,\tilde{\theta}(n))\tilde{C}(1,2;n)\tilde{C}'(2,2;n)A'(2;n,\tilde{\theta}(n)) .$$

In 7(3), we used the matrix equality $I(\theta) = C(\theta)C'(\theta)$ in partitioned form to show that $C(2,2;\theta)C'(2,2;\theta) = I(2,2;\theta)$. The same equality shows that $C(1,2;\theta)C'(2,2;\theta) = I(1,2;\theta)$, and that $C(1,1;\theta)C'(1,1;\theta) + C(1,2;\theta)C'(1,2;\theta) = I(1,1;\theta)$. The matrix equation $B(\theta) = D'(\theta)D(\theta)$ in partitioned form shows that $B(1,1;\theta) = D'(1,1;\theta)D(1,1;\theta)$, and the matrix equation $C(\theta)D(\theta) = Q(m)$ in partitioned form shows that $D^{-1}(1,1;\theta) = C(1,1;\theta)$. It follows that $C(1,2;\theta)C'(1,2;\theta) = I(1,1;\theta) - B^{-1}(1,1;\theta)$. We collect the relationships we will use later:

$$C(1,2;\theta)C'(1,2;\theta) = I(1,1;\theta) - B^{-1}(1,1;\theta)$$

$$C(2,2;\theta)C'(2,2;\theta) = I(2,2;\theta) \tag{7.20}$$

$$C(1,2;\theta)C'(2,2;\theta) = I(1,2;\theta) \ .$$

The same calculations give

$$\tilde{C}(2,2;n)\tilde{C}'(2,2;n) = \tilde{I}(2,2;n)$$

$$\tilde{C}(1,2;n)\tilde{C}'(1,2;n) = \tilde{I}(1,1;n) - \tilde{B}^{-1}(1,1;n) \tag{7.21}$$

$$\tilde{C}(1,2;n)\tilde{C}'(2,2;n) = \tilde{I}(1,2;n) \ .$$

For any square matrix M with elements M_{ij}, let $d(M)$ denote its determinant, and let $|M|$ denote $\max_{i,j} |M_{ij}|$. Define the matrix $\varepsilon(\tilde{\theta}(n),\theta^0,n)$ by the matrix equation

$$\tilde{B}(n) = B(\theta^0) + \varepsilon(\tilde{\theta}(n),\theta^0,n) \ . \tag{7.22}$$

Using (7.2), (7.16), and Bonferroni's inequality, it follows that there exist null sequences $\{\Delta_9(n,\theta^0)\}$, $\{\Delta_{10}(n,\theta^0)\}$ such that

$$P_{\theta(n)}\Big[\sum_{i=1}^{m}\sum_{j=1}^{m} M_i^*(n,\theta^0)M_j^*(n,\theta^0)|\varepsilon_{ij}(\tilde{\theta}(n),\theta^0,n)| \leq \Delta_9(n,\theta^0)\Big] \tag{7.23}$$

$$\geq 1 - \Delta_{10}(n,\theta^0) \ .$$

From (7.23) it follows that $|\varepsilon(\tilde{\theta}(n),\theta^0,n)|$ converges stochastically to zero as n increases.

Using the definition of determinant, it follows from (7.22) that when $|\varepsilon(\tilde{\theta}(n),\theta^0,n)| < 1$,

(7.24) $\quad d(\tilde{B}(n)) =$

$$d(B(\theta^0)) + \lambda m! 2^{m-1} |\varepsilon(\tilde{\theta}(n),\theta^0,n)| (1+|B(\theta^0)|)^{m-1}$$

where $|\lambda| \leq 1$. A similar formula holds for cofactors of $\tilde{B}(n)$. Using the formula for the inverse in terms of the cofactors, and (7.23), it follows that there exist matrices $\Lambda_1(n)$, $\Lambda_2(n)$, $\Lambda_3(n)$, a finite constant $\sigma(\theta^0)$, and a null sequence $\{\Delta_{11}(n,\theta^0)\}$, such that

(7.25)

$$\tilde{I}(1,1;n) - \tilde{B}^{-1}(1,1;n) = I(1,1;\theta^0) - B^{-1}(1,1;\theta^0) + |\varepsilon(\tilde{\theta}(n),\theta^0,n)|\Lambda_1(n)$$

$$\tilde{I}(2,2;n) = I(2,2;\theta^0) + |\varepsilon(\tilde{\theta}(n),\theta^0,n)|\Lambda_2(n)$$

$$\tilde{I}(1,2;n) = I(1,2;\theta^0) + |\varepsilon(\tilde{\theta}(n),\theta^0,n)|\Lambda_3(n) , \quad \text{and}$$

$$P_{\theta(n)}[|\Lambda_i(n)| < \sigma(\theta^0);\ i=1,2,3] \geq 1 - \Delta_{11}(n,\theta^0) .$$

For $i = 1,\ldots,m$, we have

(7.26)
$$A_i(n,\tilde{\theta}(n)) = A_i(n,\theta^0) - \sum_{j=1}^{m} k_j(n)(\tilde{\theta}_j(n) - \theta_j^0)B_{ji}(\theta^0)$$
$$- \sum_{j=1}^{m} k_j(n)(\tilde{\theta}_j(n) - \theta_j^0)\varepsilon_{ji}(\tilde{\theta}^{(i)}(n),\theta^0,n)$$

where, because of (7.16) and our assumption about $\theta(n)$, we have

(7.27)

$$P_{\theta(n)}[\tilde{\theta}^{(i)}(n) \text{ is in } N_n(\theta^0) \text{ for } i = 1,\ldots,m)] \geq 1 - \Delta_{12}(n,\theta^0) ,$$

for some null sequence $\{\Delta_{12}(n,\theta^0)\}$.

Let $\bar{\varepsilon}(n)$ denote the m by m matrix with $\varepsilon_{ji}(\tilde{\theta}^{(i)}(n),\theta^0,n)$ in row j and column i. Let $\bar{w}(n) = (\bar{w}_1(n),\ldots,\bar{w}_m(n))$ denote the one by m vector with $\bar{w}_i(n) = k_i(n)(\tilde{\theta}_i(n)-\theta_i^0)$ as i^{th} element. We note that $\bar{w}_i(n) = 0$ for $i = s+1,\ldots,m$. We partition $\bar{w}(n)$ as $(\bar{w}(1,n),\bar{w}(2,n))$, where $\bar{w}(1,n)$ is one by s, and $\bar{w}(2,n)$ consists of zeroes. We partition $\bar{\varepsilon}(n)$ as $\begin{bmatrix} \bar{\varepsilon}(1,1;n) & \bar{\varepsilon}(1,2;n) \\ \bar{\varepsilon}(2,1;n) & \bar{\varepsilon}(2,2;n) \end{bmatrix}$, where the northwest corner is s by s. The m equalities in (7.26) can be written as the single matrix equality.

(7.28)

$$(A(1;n,\tilde{\theta}(n)),A(2;n,\tilde{\theta}(n))) = (A(1;n,\theta^0),A(2;n,\theta^0))$$
$$-\ (\bar{w}(1,n)B(1,1;\theta^0),\bar{w}(1,n)B(1,2;\theta^0))$$
$$-\ (\bar{w}(1,n)\bar{\varepsilon}(1,1;n),\bar{w}(1,n)\bar{\varepsilon}(1,2;n)) \ .$$

Using (7.21), (7.25), and (7.28) in (7.19), we find that $\sum_{i=s+1}^{m} \tilde{Z}_i^2(n)$ can be written as the sum of the following six expressions:

(7.29) $$\sum_{i=s+1}^{m} Z_i^2(n;\theta^0)$$

(7.30)

$$-2A(1;n,\theta^0)[I(1,1;\theta^0)B(1,1;\theta^0) + I(1,2;\theta^0)B(2,1;\theta^0)-Q(s)]\bar{w}'(1,n)$$
$$-2A(2;n,\theta^0)[I(2,1;\theta^0)B(1,1;\theta^0) + I(2,2;\theta^0)B(2,1;\theta^0)]\bar{w}'(1,n)$$

(7.31)

$$\bar{w}(1,n)\begin{bmatrix} B(1,1;\theta^0)\{I(1,1;\theta^0) - B^{-1}(1,1;\theta^0)\}B(1,1;\theta^0) \\ + B(1,2;\theta^0)I(2,2;\theta^0)B'(1,2;\theta^0) \\ + 2B(1,1;\theta^0)I(1,2;\theta^0)B'(1,2;\theta^0) \end{bmatrix}\bar{w}'(1,n)$$

(7.32)

$$|\varepsilon(\hat{\theta}(n),\theta^0,n)|\begin{bmatrix} [A(1;n,\theta^0)-\bar{w}(1,n)B(1,1;\theta^0)-\bar{w}(1,n)\bar{\varepsilon}(1,1;n)]\Lambda_1(n) \\ \otimes [A(1;n,\theta^0)-\bar{w}(1,n)B(1,1;\theta^0)-\bar{w}(2,n)\bar{\varepsilon}(1,1;n)]' \\ + [A(2;n,\theta^0)-\bar{w}(1,n)B(1,2;\theta^0)-\bar{w}(1,n)\bar{\varepsilon}(1,2;n)]\Lambda_2(n) \\ \otimes [A(2;n,\theta^0)-\bar{w}(1,n)B(1,2;\theta^0)-\bar{w}(1,n)\bar{\varepsilon}(1,2;n)]' \\ + 2[A(1;n,\theta^0)-\bar{w}(1,n)B(1,1;\theta^0)-\bar{w}(1,n)\bar{\varepsilon}(1,1;n)]\Lambda_3(n) \\ \otimes [A(2;n,\theta^0)-\bar{w}(1,n)B(1,2;\theta^0)-\bar{w}(1,n)\bar{\varepsilon}(1,2;n)]' \end{bmatrix}$$

(7.33)

$$-2[A(1;n,\theta^0)-\bar{w}(1,n)B(1,1;\theta^0)]\{I(1,1;\theta^0)-B^{-1}(1,1;\theta^0)\}[\bar{w}(1,n)\bar{\varepsilon}(1,1;n)]$$

$$-2[A(2;n,\theta^0)-\bar{w}(1,n)B(1,2;\theta^0)]I(2,2;\theta^0)[\bar{w}(1,n)\bar{\varepsilon}(1,2;n)]'$$

$$-2[A(1;n,\theta^0)-\bar{w}(1,n)B(1,1;\theta^0)]I(1,2;\theta^0)[\bar{w}(1,n)\bar{\varepsilon}(1,2;n)]'$$

$$-2\bar{w}(1,n)\bar{\varepsilon}(1,1;n)I(1,2;\theta^0)[A(2;n,\theta^0)-\bar{w}(1,n)B(1,2;\theta^0)]'$$

(7.34)

$$\bar{w}(1,n)\bar{\varepsilon}(1,1;n)\{I(1,1;\theta^0)-B^{-1}(1,1;\theta^0)\}[\bar{w}(1,n)\bar{\varepsilon}(1,1;n)]'$$
$$+\ \bar{w}(1,n)\bar{\varepsilon}(1,2;n)I(2,2;\theta^0)[\bar{w}(1,n)\bar{\varepsilon}(1,2;n)]'$$
$$+\ 2\bar{w}(1,n)\bar{\varepsilon}(1,1;n)I(1,2;\theta^0)[\bar{w}(1,n)\bar{\varepsilon}(1,2;n)]'$$

Writing $I(\theta^0)B(\theta^0) = Q(m)$ and $B(\theta^0)I(\theta^0) = Q(m)$ in partitioned form and equating corresponding elements, it is easily shown that the matrices in square brackets in expressions (7.30) and (7.31) have all elements equal to zero, so expressions (7.30) and (7.31) are equal to zero.

Define the vector $Y(n,\theta^0)$ as $A(n,\theta^0)C(\theta^0)$ - $(k_1(n)(\theta_1(n)-\theta_1^0),\ \ldots,\ k_m(n)(\theta_m(n)-\theta_m^0))[C'(\theta^0)]^{-1}$. It follows from Theorem 7.2 that the asymptotic joint distribution of the components of $Y(n,\theta^0)$ is that of m independent standard normal variables, and the approach to this limiting distribution is uniform over the class of $\theta(n)$ we are considering. Using (7.16), (7.27), (7.25) and (7.2), it follows that we can write $\sum_{i=s+1}^{m} \tilde{Z}_i^2(n)$ as $\sum_{i=s+1}^{m} Z_i^2(n,\theta^0) + \bar{R}(n)$, where

(7.35) $$P[|\bar{R}(n)| < \Delta_{13}(n,\theta^0)] > 1 - \Delta_{14}(n,\theta^0)$$

for some null sequences $\{\Delta_{13}(n,\theta^0)\}$, $\{\Delta_{14}(n,\theta^0)\}$.

The event $\{\sum_{i=s+1}^{m} \tilde{Z}_i^2(n) > T(\alpha)\}$ is implied by the event $\{\sum_{i=s+1}^{m} Z_i^2(n,\theta^0) > T(\alpha) + \Delta_{13}(n,\theta^0)$ and $|\bar{R}(n)| < \Delta_{13}(n,\theta^0)\}$. Also,

$$\Big| P\Big[\sum_{i=s+1}^{m} Z_i^2(n,\theta^0) > T(\alpha) + \Delta_{13}(n,\theta^0) \quad \text{and} \quad |\overline{R}(n)| < \Delta_{13}(n,\theta^0)\Big] - P\Big[\sum_{i=s+1}^{m} Z_i^2(n,\theta^0) > T(\alpha) + \Delta_{13}(n,\theta^0)\Big] \Big| < \Delta_{14}(n,\theta^0)$$

so that

$$P\Big[\sum_{i=s+1}^{m} \tilde{Z}_i^2(n) > T(\alpha)\Big] \geqq P\Big[\sum_{i=s+1}^{m} Z_i^2(n,\theta^0) > T(\alpha) + \Delta_{13}(n,\theta^0)\Big] - \Delta_{14}(n,\theta^0) .$$

Similarly, we find

$$P\Big[\sum_{i=s+1}^{m} \tilde{Z}_i^2(n) \leqq T(\alpha)\Big] \geqq P\Big[\sum_{i=s+1}^{m} Z_i^2(n,\theta^0) \leqq T(\alpha) - \Delta_{13}(n,\theta^0)\Big] - \Delta_{14}(n,\theta^0) ,$$

so we get

$$(7.36) \qquad P\Big[\sum_{i=s+1}^{m} Z_i^2(n,\theta^0) > T(\alpha) + \Delta_{13}(n,\theta^0)\Big] - \Delta_{14}(n,\theta^0) \leqq P\Big[\sum_{i=s+1}^{m} \tilde{Z}_i^2(n) > T(\alpha)\Big] \leqq P\Big[\sum_{i=s+1}^{m} Z_i^2(n,\theta^0) > T(\alpha) - \Delta_{13}(n,\theta^0)\Big] + \Delta_{14}(n,\theta^0)$$

It follows from Theorem 7.4 that

$$\left| P\left[\sum_{i=s+1}^{m} Z_i^2(n,\theta^0) > T(\alpha) + \Delta_{13}(n,\theta^0)\right] - P\left[\sum_{i=s+1}^{m} Z_i^2(n,\theta^0) > T(\alpha) - \Delta_{13}(n,\theta^0)\right] \right| < \Delta_{15}(n,\theta^0)$$

and

$$\left| P\left[\sum_{i=s+1}^{m} \tilde{Z}_i^2(n,\theta^0) > T(\alpha) + \Delta_{13}(n,\theta^0)\right] - \left[1 - \chi_{m-s}(T(\alpha);\delta^{1/2})\right] \right| < \Delta_{16}(n,\theta^0)$$

for some null sequences $\{\Delta_{15}(n,\theta^0)\}$, $\{\Delta_{16}(n,\theta^0)\}$. It follows that

$$\left| P\left[\sum_{i=s+1}^{m} \tilde{Z}_i^2(n) > T(\alpha)\right] - \left[1 - \chi_{m-s}(T(\alpha);\delta^{1/2})\right] \right| \leq \Delta_{17}(n,\theta^0)$$

for some null sequence $\{\Delta_{17}(n,\theta^0)\}$. This shows that the test $W(n)$ has the same asymptotic properties as the test T_n^* described in (7.15). Thus $W(n)$ is asymptotically optimal for the nonartificial problem.

7(5). In this section we give two examples. The first example is as follows.

For each n, $Y_1(n), \ldots, Y_n(n)$ are independent and identically distributed random variables, each with possible values $0,1,2,\ldots$. The hypothesis to be tested is that the common distribution is Poisson, with an unspecified parameter. Let $X_i(n)$ denote the number of values $(Y_1(n),\ldots,Y_n(n))$ equal to i, for $i = 0,\ldots,m-1$, and let $X_m(n)$ denote $n - \sum_{i=1}^{m-1} X_i(n)$. Our decision is to be based on $X(n) = (X_0(n),\ldots,X_m(n))$. Let p_i denote the unknown probability that $Y_1(n) = i$, for $i = 0,\ldots,m-1$. Let p_m denote $1 - \sum_{i=0}^{m-1} p_i$. We assume $p_i > 0$ for $i = 0,\ldots,m$. The hypothesis to be tested is that $p_i = \frac{\lambda^i e^{-\lambda}}{i!}$ for $i = 0,\ldots,m-1$, where λ is some unspecified positive value.

We reparametrize this problem by defining parameters $\theta_1, \ldots, \theta_m$ as follows:

$$\theta_1 = -\log p_0$$

$$\theta_i = \frac{\theta_1^{i-1} e^{-\theta_1}}{(i-1)!} - p_{i-1}\ , \quad i = 2,\ldots,m.$$

There is a one to one correspondence between $\theta \equiv (\theta_1,\ldots,\theta_m)$ and $(p_0,\ldots,p_{m-1})$. In terms of θ, the hypothesis to be tested is that $\theta_2 = 0,\ldots,\theta_m = 0$, with θ_1 some unspecified value in the open interval $(0,\infty)$. The positivity assumptions about p_i give the following Θ:

$$0 < \theta_1 < \infty$$

$$\theta_i < \frac{\theta_1^{i-1} e^{-\theta_1}}{(i-1)!} \quad \text{for } i = 2,\ldots,m$$

$$\sum_{i=0}^{m-1} \frac{\theta_1^i e^{-\theta_1}}{i!} < 1 + \sum_{i=2}^{m} \theta_i\ .$$

We note that Θ contains the point $(\theta_1,0,\ldots,0)$ for any finite positive θ_1.

$$K_n(X(n)|\theta_1,\ldots,\theta_m) = n! \prod_{i=0}^{m} \frac{[p_i(\theta)]^{X_i(n)}}{(X_i(n))!}$$

where $p_i(\theta)$ is given by the equations defining θ above for $i = 0,\ldots,m-1$, and $p_m(\theta)$ is defined as $1 - \sum_{i=0}^{m-1} p_i(\theta)$. Then we have

$$-\frac{1}{n}\frac{\partial^2 \log K_n(X(n)|\theta)}{\partial\theta_i\partial\theta_j} = \frac{X_m(n)}{n\, p_m^2(\theta)} \quad \text{if } i\neq j,\ i>1,\ j>1;$$

$$-\frac{1}{n}\frac{\partial^2 \log K_n(X(n)|\theta)}{\partial\theta_i^2} = \frac{X_{i-1}(n)}{n\, p_{i-1}^2(\theta)} + \frac{X_m(n)}{n\, p_m^2(\theta)} \quad \text{if } i>1;$$

$$-\frac{1}{n}\frac{\partial^2 \log K_n(X(n)|\theta)}{\partial\theta_1\partial\theta_i} = -\frac{X_{i-1}(n)}{n\, p_{i-1}^2(\theta)} e^{-\theta_1}\left[\frac{(i-1)\theta_1^{i-2}-\theta_1^{i-1}}{(i-1)!}\right]$$

$$-\frac{X_m(n)}{n\, p_m^2(\theta)} e^{-\theta_1}\sum_{j=0}^{m-1}\left[\frac{j\theta_1^{j-1}-\theta_1^{j}}{j!}\right] \quad \text{if } i>1;$$

$$-\frac{1}{n}\frac{\partial^2 \log K_n(X(n)|\theta)}{\partial\theta_1^2} = -\sum_{j=0}^{m-1}\frac{X_j(n)}{n\, p_j(\theta)}\frac{\partial^2 p_j(\theta)}{\partial\theta_1^2}$$

$$+\sum_{j=0}^{m-1}\frac{X_j(n)}{n\, p_j^2(\theta)}\left[\frac{e^{-\theta_1}(j\theta_1^{j-1}-\theta_1^{j})}{j!}\right]^2 + \frac{X_m(n)}{n\, p_m(\theta)}\sum_{j=0}^{m-1}\frac{\partial^2 p_j(\theta)}{\partial\theta_1^2}$$

$$+\frac{X_m(n)}{n\, p_m^2(\theta)}\left[\sum_{j=0}^{m-1}\frac{\partial p_j(\theta)}{\partial\theta_1}\right]^2 .$$

Since $\frac{X_j(n)}{n}$ converges stochastically to $p_j(\theta)$ as n increases, the first part of assumption (7.1) is true in this example, with $k_i(n) = \sqrt{n}$ for $i = 1,\ldots,m$, and $B_{ij}(\theta)$ given by replacing $\frac{X_j(n)}{n}$ by $p_j(\theta)$ in the expressions for $-\frac{1}{n}\frac{\partial^2 \log K_n(X(n)|\theta)}{\partial\theta_i \partial\theta_j}$ given above. The continuity of $B_{ij}(\theta)$ follows directly. The only part of assumption (7.1) remaining to be verified is the nonsingularity of $B(\theta)$. This is verified by the following argument. If we did not reparametrize, but kept the original parameters $p_0, \ldots, p_{m-1}$, it is easily seen that the corresponding B matrix is nonsingular. Since the transformation from $p_0, \ldots, p_{m-1}$ to $\theta_1, \ldots, \theta_m$ is nonsingular, $B(\theta)$ must be nonsingular.

Next we verify that assumption (7.2) holds, with $M_i^*(n,\theta) = n^{\frac{1}{6}-\delta}$, where δ is any fixed value in the open interval $(0,\frac{1}{6})$. For if we change each θ_k by no more than $\frac{n^{\frac{1}{6}-\delta}}{n^{1/2}}$, the effect on $-\frac{1}{n}\frac{\partial^2 \log K_n(X(n)|\theta)}{\partial\theta_i \partial\theta_j}$ is of the order of $\frac{n^{\frac{1}{6}-\delta}}{n^{1/2}}$, and thus

$$\sup_{\theta \text{ in } N_n(\theta^0)} |\varepsilon_{ij}(\theta,\theta^0,n)| \text{ is of the order of } \frac{n^{\frac{1}{6}-\delta}}{n^{1/2}}, \text{ and}$$

$$\sum_{i=1}^{m}\sum_{j=1}^{m}(n^{\frac{1}{6}-\delta})^2 \sup_{\theta \text{ in } N_n(\theta^0)} |\varepsilon_{ij}(\theta,\theta^0,n)| \text{ is of the order of}$$

$\frac{(n^{\frac{1}{6}-\delta})^3}{n^{1/2}} = n^{-3\delta}$, which clearly implies that assumption (7.2) holds.

For $\gamma_1(n)$ of section 7(4), we can use $-\log\left[\frac{X_0(n)}{n}\right]$. This completes the discussion of the first example.

For our second example, we have an S-state stationary Markov chain, with transition matrix $P(\theta)$, where the element in row i

and column j of $P(\theta)$ is θ_{ij}, where $\theta_{iS} = 1 - \theta_{i1} - \cdots - \theta_{i,S-1}$ for $i = 1,\ldots,S$. Thus there are $S(S-1)$ independent parameters, θ_{ij} for $i = 1,\ldots,S$, $j = 1,\ldots,S-1$. Θ consists of the set of parameters such that θ_{ij} is in the open interval $(0,1)$ for $i = 1,\ldots,S$, $j = 1,\ldots,S$. The problem is to test the hypothesis that some subset of the independent parameters has specified values. $X_0 = 1$, and $X_0, X_1, \ldots, X_n$ are the observed states. Define N_{ij} as the number of transitions from state i to state j in the sequence $X_0, X_1, \ldots, X_n$, for $i = 1,\ldots,S$, $j = 1,\ldots,S$. Then $K_n(X(n)|\theta) = \prod_{i=1}^{S} \prod_{j=1}^{S} \theta_{ij}^{N_{ij}}$, and thus

$$-\frac{1}{n}\frac{\partial^2 \log K_n(X(n)|\theta)}{\partial\theta_{ij}\partial\theta_{i'j'}} = 0 \quad \text{if} \quad i \neq i' ,$$

$$-\frac{1}{n}\frac{\partial^2 \log K_n(X(n)|\theta)}{\partial\theta_{ij}\partial\theta_{ij'}} = \frac{N_{iS}}{n\theta_{iS}^2} \quad \text{if} \quad j \neq j' ,$$

$$-\frac{1}{n}\frac{\partial^2 \log K_n(X(n)|\theta)}{\partial\theta_{ij}^2} = \frac{N_{ij}}{n\theta_{ij}^2} + \frac{N_{iS}}{n\theta_{iS}^2}$$

for $i = 1,\ldots,S$, $j = 1,\ldots,S-1$.

Let $(V_1(\theta),\ldots,V_S(\theta)) \equiv V(\theta)$ denote the vector of stationary probabilities for the chain. Then $\frac{N_{ij}}{n}$ converges stochastically to $V_i(\theta)\theta_{ij}$ as n increases. The rest of the analysis is similar to the analysis of the first example. If θ_{ij} is one of the nuisance parameters to be estimated, then $\tilde{\theta}_{ij}(n)$ can be taken as $\frac{N_{ij}}{N_{i1}+\cdots+N_{iS}}$.

APPENDIX

The purpose of this appendix is to illustrate the application of the theory to various types of cases. The first part of the appendix gives sufficient regularity conditions for the theory of m.ℓ. estimators described in Section 6(1) to hold in certain cases where the components of $X(n)$ are not necessarily independent and identically distributed. The second part of the appendix gives some examples where the regularity conditions are violated, but m.p. theory can still be applied.

In several of the examples to be discussed, the dimension m of Θ is greater than one, and each component of the vector of parameters requires its own normalizing factor. $k_i(n)$ will be the symbol used for the normalizing factor for the parameter θ_i $(i = 1,\ldots,m)$. This differs from the notation used in the rest of the monograph, where the discussion was carried out in detail only for the case $m = 1$.

PART I

In this first part of the appendix, we assume the following:

A(1). There exist m sequences of nonrandom positive quantities, $\{k_1(n)\}, \ldots, \{k_m(n)\}$, with $\lim_{n\to\infty} k_i(n) = \infty$ for $i = 1,\ldots,m$, such that for any $\theta^0 = (\theta_1^0,\ldots,\theta_m^0)$ in Θ,

$$-\frac{1}{k_i(n)k_j(n)} \frac{\partial^2}{\partial\theta_i\partial\theta_j} \log K_n(X(n)|\theta_1,\ldots,\theta_m)]_{\theta^0}$$

converges stochastically as n increases to a nonrandom quantity, say $B_{ij}(\theta^0)$, when θ^0 is the true parameter value, for $i,j = 1,\ldots,m$. $B_{ij}(\theta^0)$ is assumed to be a continuous function of θ^0. Let $B(\theta^0)$ denote the m by m matrix with $B_{ij}(\theta^0)$ in row i and column j. We assume $[B(\theta^0)]^{-1}$ exists, and denote it by $I(\theta^0)$.

A(2). For each θ^0 in Θ, we assume there exist m sequences of nonrandom positive quantities $\{M_1^*(n,\theta^0)\}, \ldots, \{M_i^*(n,\theta^0)\}, \ldots, \{M_m^*(n,\theta^0)\}$, satisfying the following conditions:

(a) $\lim_{n\to\infty} M_i^*(n,\theta^0) = \infty$, $i = 1,\ldots,m$.

(b) $\lim_{n\to\infty} \frac{M_i^*(n,\theta^0)}{k_i(n)} = 0$, $i = 1,\ldots,m$.

(c) Let $N_n(\theta^0)$ denote the set of all vectors $\theta = (\theta_1,\ldots,\theta_k)$ such that $|\theta_i - \theta_i^0| \leq \frac{M_i^*(n,\theta^0)}{k_i(n)}$ for $i = 1,\ldots,m$. (Note that for all sufficiently large n, $N_n(\theta^0)$ is contained in Θ.) We denote

$$-\frac{1}{k_i(n)k_j(n)} \frac{\partial^2}{\partial\theta_i\partial\theta_j} \log K_n(X(n)|\theta) - B_{ij}(\theta^0)$$

by $\varepsilon_{ij}(\theta,\theta^0,n)$. For any $\gamma > 0$, let $S_n(\theta^0,\gamma)$ denote the region in $X(n)$-space where

$\sum_{i=1}^{m} \sum_{j=1}^{m} M_i^*(n,\theta^0)M_j^*(n,\theta^0) \sup_{\theta \text{ in } N_n(\theta^0)} |\varepsilon_{ij}(\theta,\theta^0,n)| < \gamma$. We assume that there exist two sequences of nonrandom positive quantities $\{\gamma(n,\theta^0)\}$, $\{\delta(n,\theta^0)\}$, with $\lim_{n\to\infty} \gamma(n,\theta^0) = 0$ and $\lim_{n\to\infty} \delta(n,\theta^0) = 0$, such that for each n and each θ in $N_n(\theta^0)$, $P_\theta[X(n) \text{ in } S_n(\theta;\gamma(n,\theta^0))] > 1 - \delta(n,\theta^0)$.

The list of assumptions is now complete. Before motivating these assumptions, we show that in the special case $m = 1$ and $X(n) = (X_1,\ldots,X_n)$ with $X_1, \ldots, X_n$ independent with common marginal density $f(x|\theta)$, so that $K_n(X(n)|\theta) = \prod_{i=1}^{n} f(X_i|\theta)$, these assumptions are much less restrictive than those given in 6(1), which are typical of the standard literature. Since $m = 1$, we drop all subscripts i,j. In our discussion, $\{\Delta_i(n,\theta^0)\}$ is, for each i, some sequence of nonrandom positive quantities depending only on n and θ^0, with $\lim_{n\to\infty} \Delta_i(n,\theta^0) = 0$. Setting $k(n) = \sqrt{n}$ and $B(\theta) = E_\theta\{\frac{\partial^2}{\partial\theta^2} \log f(X_1|\theta)\}$, our assumption A(1) is seen to hold, with the positivity of $B(\theta^0)$ following from (6.5)(c) and the continuity of $B(\theta^0)$ following from (6.8). From (6.6)(b) we get

(A.1.1) For each n and each θ in $N_n(\theta^0)$,

$$P\left[\left| -\frac{1}{n} \frac{\partial^2}{\partial\theta^2} \log K_n(X(n)|\theta) - B(\theta) \right| < \Delta_1(n,\theta^0)\right]$$

$$> 1 - \Delta_2(n,\theta^0) .$$

From the continuity of $B(\theta)$, we get

(A.1.2) For each n and each θ in $N_n(\theta^0)$, $|B(\theta) - B(\theta^0)|$

$$< \Delta_3(n,\theta^0) .$$

From (6.8) we get that for all sufficiently large n,

(A.1.3)

$$\sup_{\theta^{(1)},\theta^{(2)} \text{ in } N_n(\theta^0)} \left\{\left|-\frac{1}{n} \frac{\partial^2}{\partial\theta^2} \log K_n(X(n)|\theta)\right]_{\theta^{(1)}} + \frac{1}{n} \frac{\partial^2}{\partial\theta^2} \log K_n(X(n)|\theta)\right]_{\theta^{(2)}}\right|\right\}$$

$$\leq k_{\theta^0} \frac{2M^*(n,\theta^0)}{\sqrt{n}}$$

From (A.1.2) and (A.1.3), for all sufficiently large n and all θ^*, θ in $N_n(\theta^0)$, we get

$$(A.1.4) \qquad |\varepsilon(\theta^*, \theta^0, n)| \leq \left| -\frac{1}{n} \frac{\partial^2}{\partial\theta^2} \log K_n(X(n)|\theta) - B(\theta) \right|$$

$$+ \Delta_3(n,\theta^0) + k_{\theta^0} \frac{2M^*(n,\theta^0)}{\sqrt{n}} .$$

From (A.1.1) and (A.1.4), we get that for all sufficiently large n, and all θ in $N_n(\theta^0)$,

$$P_\theta \left[(M^*(n,\theta^0))^2 \sup_{\theta^* \text{ in } N_n(\theta^0)} |\varepsilon(\theta^*,\theta^0,n)| \leq (M^*(n,\theta^0))^2 \left[\Delta_1(n,\theta^0) + \Delta_3(n,\theta^0) + \frac{2k_{\theta^0} M^*(n,\theta^0)}{\sqrt{n}} \right] \right]$$

$$> 1 - \Delta_2(n,\theta^0) .$$

Now if we define $M^*(n,\theta^0)$ as

$$\min \left[\left[\frac{1}{\Delta_1(n,\theta^0)} \right]^{1/4}, \left[\frac{1}{\Delta_3(n,\theta^0)} \right]^{1/4}, n^{1/12} \right]$$

it is easy to verify that assumption A(2) is satisfied.

The motivation for assumptions A(1) and A(2) is fairly obvious. The purpose of A(1) is to guarantee (asymptotically) that $K_n(X(n)|\theta)$ will have a peak near the true value θ^0. The assumption A(2) guarantees that a small change in θ does not lead to a large change in the asymptotic behavior of $K_n(X(n)|\theta)$.

What conclusions follow from assumptions A(1) and A(2)? It was shown in [10] and [11] that if A(1) and A(2) hold, the following hold:

(1) Under any θ^0 in Θ, and for each n, there is a neighborhood $C_n(\theta^0)$ of θ^0, with diameter approaching zero as n increases, such that $\lim_{n\to\infty} P_{\theta^0}[K_n(X(n)|\theta)$ has a relative maximum w.r.t. θ in $C_n(\theta^0)] = 1$. If for each n, $\hat{\theta}(n) = (\hat{\theta}_1(n),\ldots,\hat{\theta}_m(n))$ is a point at which such a relative maximum occurs, then $k_1(n)(\hat{\theta}_1(n)-\hat{\theta}_1^0)$, ..., $k_m(n)(\hat{\theta}_m(n)-\theta_m^0)$ have asymptotically an m-variate normal distribution with zero means and covariance matrix $I(\theta^0)$.

(2) $\hat{\theta}(n)$ is an m.p. estimator with respect to any measurable convex R which is symmetric about the origin.

(3) Suppose for each n we have available a vector $\bar{\theta}(n) = (\bar{\theta}_1(n),\ldots,\bar{\theta}_m(n))$ such that for any sequence $\{L(n)\}$ of positive nonrandom quantities with $\lim_{n\to\infty} L(n) = \infty$, $\lim_{n\to\infty} P_{\theta^0}[k_i(n)|\bar{\theta}_i(n) - \theta_i^0| < L(n);\ i = 1,\ldots,m] = 1$, for each θ^0 in Θ. Define the vector $A(n;\bar{\theta}(n))$ as the row vector with i^{th} element given by $\frac{1}{k_i(n)} \frac{\partial}{\partial\theta_i} \log K_n(X(n)|\theta]_{\bar{\theta}(n)}$. Define the vector $\hat{\theta}^*(n)$ by the matrix equation $(k_1(n)(\hat{\theta}_1^*(n) - \bar{\theta}_1(n)), \ldots, k_m(n)(\hat{\theta}_m^*(n) - \bar{\theta}_m(n))) = A(n;\bar{\theta}(n))I(\bar{\theta}(n))$. Then $\hat{\theta}^*(n)$ has the same asymptotic distribution as $\hat{\theta}(n)$, and is therefore an m.p. estimator. Note that here $\bar{\theta}(n)$ is a "preliminary estimator" which does not have to be asymptotically efficient.

We illustrate with five examples.

Example 1. For each n, X_1', ..., X_n' are independent and identically distributed, with common density function $f(x-\theta)$, common distribution function $F(x-\theta)$, where f is a known function and θ is an unknown location parameter. Thus $m = 1$ in this example, and we dispense with subscripts. We assume f is positive everywhere, and has a continuous second derivative everywhere. p, q are given values with $0 < p < q < 1$. $X_1 < \cdots < X_n$ are the ordered values of $X_1', \ldots, X_n'$. The estimation of θ is to be based only on $X_{[np]}, \ldots, X_{[nq]}$. Thus the vector $X(n)$ is $(X_{[np]}, \ldots, X_{[nq]})$, and $K_n(X(n)|\theta)$ is given by

$$\frac{n!}{([np]-1)!(n-[nq])!}(F(X_{[np]}-\theta))^{[np]-1}(1-F(X_{[nq]}-\theta))^{n-[nq]}\prod_{i=[np]}^{[nq]} f(X_i-\theta)$$

for $X_{[np]} < \cdots < X_{[nq]}$, and $K_n(X(n)|\theta)$ is zero otherwise. In this example, for any $\Delta > 0$, $n^{1/2-\Delta} \max_{[np]\leq i\leq[nq]} |X_i-\theta^0-F^{-1}(\frac{i}{n})|$ converges stochastically to zero as n increases. From this it follows easily that assumptions A(1) and A(2) are satisfied with

$$k(n) = \sqrt{n}, \quad \text{and} \quad B(\theta^0) = \frac{f^2(F^{-1}(p))}{p} + \frac{f^2(F^{-1}(q))}{1-q}$$

$$+ \int_{F^{-1}(p)!}^{F^{-1}(q)} \left[\frac{f'(y)}{f(y)}\right]^2 f(y)dy.$$ Since $\sqrt{n}\,(X_{[np]} - \theta^0 - F^{-1}(p))$ has asymptotically a normal distribution with mean zero and finite variance, we can use $X_{[np]} - F^{-1}(p)$ as $\bar{\theta}(n)$. We note that $F^{-1}(p)$ is known.

In the paper [16], an estimation problem where f is known to be symmetric and to have f'' satisfy a Lipschitz condition, but is otherwise unknown, was solved by estimating f and then using the estimate of f to construct the estimate of θ.

Example 2. This example modifies example 1 by introducing

a scale parameter, so the common density is $\frac{1}{\theta_2} f\left(\frac{x-\theta_1}{\theta_2}\right)$, with f known, θ_1 and $\theta_2 > 0$ unknown. Here $m = 2$, and $K_n(X(n)|\theta) =$

$$\frac{n!}{([np]-1)!(n-[nq])!} \left[F\left(\frac{X_{[np]}-\theta_1}{\theta_2}\right)\right]^{[np]-1}$$

$$\left[1-F\left(\frac{X_{[nq]}-\theta_1}{\theta_2}\right)\right]^{n-[nq]} \prod_{i=[np]}^{[nq]} \left[\frac{1}{\theta_2} f\left(\frac{X_i-\theta_1}{\theta_2}\right)\right]$$

if $X_{[np]} < \cdots < X_{[nq]}$. Using the fact that for any $\Delta > 0$,

$n^{1/2-\Delta} \max\limits_{[np]\leq i\leq[nq]} \left|\frac{X_i-\theta_1}{\theta_2} - F^{-1}\left(\frac{i}{n}\right)\right|$ converges stochastically to zero as n increases, we can proceed as in example 1. We find $k_1(n) = k_2(n) = \sqrt{n}$, and

$$B_{11}(\theta^0) = \left[\frac{1}{\theta_2^0}\right]^2 \left[\frac{f^2(F^{-1}(p))}{p} + \frac{f^2(F^{-1}(q))}{1-q} + \int_{F^{-1}(p)}^{F^{-1}(q)} \left(\frac{f'(y)}{f(y)}\right)^2 f(y)dy\right],$$

$$B_{12}(\theta^0) = \left[\frac{1}{\theta_2^0}\right]^2 \left[\int_{F^{-1}(p)}^{F^{-1}(q)} y \left(\frac{f'(y)}{f(y)}\right)^2 f(y)dy + f(F^{-1}(q)) - f(F^{-1}(p)) + F^{-1}(p)\,\frac{f^2(F^{-1}(p))}{p} + F^{-1}(q)\,\frac{f^2(F^{-1}(q))}{1-q}\right]$$

$$B_{22}(\theta^0) = \left[\frac{1}{\theta_2^0}\right]^2 \left[\frac{(F^{-1}(p))^2 f^2(F^{-1}(p))}{p} + \frac{(F^{-1}(q))^2 f^2(F^{-1}(q))}{1-q} - q+p + \int_{F^{-1}(p)}^{F^{-1}(q)} \left(y\,\frac{f'(y)}{f(y)}\right)^2 f(y)dy + 2F^{-1}(q)f(F^{-1}(q)) - 2F^{-1}(p))f(F^{-1}(p))\right]$$

In this example, $\sqrt{n}\left[\frac{X_{[np]}-\theta_1^0}{\theta_2^0} - F^{-1}(p)\right]$, $\sqrt{n}\left[\frac{X_{[nq]}-\theta_1^0}{\theta_2^0} - F^{-1}(q)\right]$ have asymptotically a joint normal distribution with zero means and finite variances. It follows that we can define $\bar{\theta}_2(n) = \frac{X_{[nq]}-X_{[np]}}{F^{-1}(q)-F^{-1}(p)}$, $\bar{\theta}_1(n) = X_{[nq]} - \bar{\theta}_2(n)F^{-1}(q)$.

Example 3. We have a 2-state stationary Markov chain, with transition matrix $\begin{bmatrix} \theta_1 & 1-\theta_1 \\ \theta_2 & 1-\theta_2 \end{bmatrix}$, $0 < \theta_1 < 1$, $0 < \theta_2 < 1$. $X_0 = 1$, and $X_0, X_1, \ldots, X_n$ are the observed states. Define N_{ij} as the number of transitions from state i to state j in the sequence $X_0, X_1, \ldots, X_n$, for $i,j = 1,2$. Then $K_n(X(n)|\theta) = \theta_1^{N_{11}}(1-\theta_1)^{N_{12}}\theta_2^{N_{21}}(1-\theta_2)^{N_{22}}$. The stationary probabilities are

$\frac{\theta_2}{1-\theta_1+\theta_2}$, $\frac{1-\theta_1}{1-\theta_1+\theta_2}$, and it follows that as n increases, $\frac{N_{11}}{n}$ converges stochastically to $\frac{\theta_1^0\theta_2^0}{1-\theta_1^0+\theta_2^0}$, $\frac{N_{12}}{n}$ converges stochastically to $\frac{(1-\theta_1^0)\theta_2^0}{1-\theta_1^0+\theta_2^0}$, $\frac{N_{21}}{n}$ converges stochastically to $\frac{\theta_2^0(1-\theta_1^0)}{1-\theta_1^0+\theta_2^0}$, and $\frac{N_{22}}{n}$ converges stochastically to $\frac{(1-\theta_1^0)(1-\theta_2^0)}{1-\theta_1^0+\theta_2^0}$. From this, it follows that our assumptions are satisfied with $k_1(n) = k_2(n) = \sqrt{n}$, and $B_{11}(\theta^0) = \frac{\theta_2^0}{\theta_1^0(1-\theta_1^0)(1-\theta_1^0+\theta_2^0)}$, $B_{12}(\theta^0) = 0$,

$B_{22}(\theta^0) = \frac{1-\theta_1^0}{\theta_2^0(1-\theta_2^0)(1-\theta_1^0+\theta_2^0)}$. $\hat{\theta}_1(n) = \frac{N_{11}}{N_{11}+N_{12}}$, $\hat{\theta}_2(n) = \frac{N_{21}}{N_{21}+N_{22}}$.

It is clear that similar results hold for an ergodic chain with any finite number of states.

Example 4. We have a 2-state stationary Markov chain, with transition matrix $\begin{bmatrix} \theta & 1-\theta \\ \sin^2\frac{\pi}{2}\theta & \cos^2\frac{\pi}{2}\theta \end{bmatrix}$, $0 < \theta < 1$. $X_0 = 1$, and $X_0, X_1, \ldots, X_n$ are the observed states. Define N_{ij} as in example 3. Then $K_n(X(n)|\theta) = \theta^{N_{11}}(1-\theta)^{N_{12}}(\sin^2 \frac{\pi}{2}\theta)^{N_{21}}(\cos^2\frac{\pi}{2}\theta)^{N_{22}}$.

The stationary probabilities are $\frac{\sin^2 \frac{\pi}{2}\theta}{1-\theta+\sin^2 \frac{\pi}{2}\theta}$, $\frac{1-\theta}{1-\theta+\sin^2 \frac{\pi}{2}\theta}$, and it follows that as n increases, $\frac{N_{11}}{n}$ converges stochastically to $\frac{\theta^0 \sin^2 \frac{\pi}{2}\theta^0}{1-\theta^0+\sin^2 \frac{\pi}{2}\theta^0}$, $\frac{N_{12}}{n}$ converges stochastically to $\frac{(1-\theta^0)\sin^2 \frac{\pi}{2}\theta^0}{1-\theta^0+\sin^2 \frac{\pi}{2}\theta^0}$, $\frac{N_{21}}{n}$ converges stochastically to $\frac{(\sin^2 \frac{\pi}{2}\theta^0)(1-\theta^0)}{1-\theta^0+\sin^2 \frac{\pi}{2}\theta^0}$, and $\frac{N_{22}}{n}$ converges stochastically to $\frac{(1-\theta^0)\cos^2 \frac{\pi}{2}\theta^0}{1-\theta^0+\sin^2 \frac{\pi}{2}\theta^0}$. From this, it follows that A(1) and A(2) are satisfied with $k(n) = \sqrt{n}$, and $B(\theta^0) =$

$$\frac{1}{1-\theta^0+\sin^2 \frac{\pi}{2}\theta^0}\left[\frac{\sin^2 \frac{\pi}{2}\theta^0}{\theta^0(1-\theta^0)} + \frac{\pi^2(1-\theta^0)}{2(\sin^2 \frac{\pi}{2}\theta^0)(\cos^2 \frac{\pi}{2}\theta^0)}\right].$$

$\overline{\theta}(n)$ can be taken as $\frac{N_{11}}{N_{11}+N_{12}}$.

It is clear that similar results apply to an ergodic chain with any finite number of states, whose transition probabilities are any reasonable functions of a finite number of parameters.

Example 5. $X_1, \ldots, X_n$ are independent, each with density function $\frac{1}{\pi\theta_2}\left[\frac{1}{1+\left(\frac{x-\theta_1}{\theta_2}\right)^2}\right]$, $\theta_2 > 0$. This Cauchy distribution is a standard case, but the computation of $\hat{\theta}_1(n)$, $\hat{\theta}_2(n)$ in closed form is impossible. The purpose of this example is to compute $\hat{\theta}_1^*(n)$, $\hat{\theta}_2^*(n)$. Here of course $k_1(n) = k_2(n) = \sqrt{n}$. A simple computation gives $B_{11}(\theta^0) = \frac{1}{2(\theta_2^0)^2}$, $B_{12}(\theta^0) = 0$, $B_{22}(\theta^0) = \frac{2}{(\theta_2^0)^2}$. Let $Z_1(n)$, $Z_2(n)$, $Z_3(n)$ denote the first, second, and third sample quartiles, respectively. $\sqrt{n}\left[\frac{Z_2(n) - \theta_1^0}{\theta_2^0}\right]$, $\sqrt{n}\left[\frac{Z_3(n) - Z_1(n)}{\theta_2^0} - 2\right]$ have asymptotically a joint normal distribution with zero means and finite variances. It follows that we can define $\bar{\theta}_1(n) = Z_2(n)$, $\bar{\theta}_2(n) = \frac{Z_3(n) - Z_1(n)}{2}$. Thus $\hat{\theta}_1^*(n)$, $\hat{\theta}_2^*(n)$ can be written in closed form.

PART II

In various parts of Chapters 5 and 6, examples were given of the application of the theory to cases not satisfying the regularity conditions of 6(1) or Part I of this Appendix. We now give some additional examples of this type.

<u>Example 6</u>. $X_1, \ldots, X_n$ are independent each with density $f(x-\theta)$, where $f(y) = 0$ if $y > 0$, $f(y)$ is continuous on the left at $y = 0$, and $f(y)$ is nondecreasing in y for all $y < 0$. $R = (-r,r)$. (Note that this example is similar to some of the examples of Chapter 5, but we have replaced the assumptions made in Chapter 5 by a monotonicity assumption about f.) It is easily verified that $K_n(X(n)|\theta) = 0$ if $\theta < \max(X_1,\ldots,X_n)$, and $K_n(X(n)|\theta)$ is nonincreasing in θ if $\theta > \max(X_1,\ldots,X_n)$. Thus $\max(X_1,\ldots,X_n) + \frac{r}{k(n)}$ is an m.p. estimator. Since $\lim_{n\to\infty} P_\theta[n\{\max(X_1,\ldots,X_n) - \theta\} \leq y] = e^{yf(0-)}$ if $y \leq 0$, 1 if $y > 0$, $k(n)$ can be taken as n.

<u>Example 7</u>. This example is a two-dimensional analogue of example 6. $(X_{11},X_{21}), \ldots, (X_{1n},X_{2n})$ are n independent pairs, each with bivariate density $f(x_1-\theta_1, x_2-\theta_2)$, where $f(y_1,y_2) = 0$ if $y_1 > 0$ or $y_2 > 0$, $f(y_1,y_2)$ is nondecreasing in each of y_1, y_2 for all $y_1 < 0$ and $y_2 < 0$, and $\lim_{\substack{y_1\to 0-\\ y_2\to 0-}} f(y_1,y_2) = f(0,0)$, $\lim_{y_1\to 0-} f_1(y_1) = f_1(0)$, $\lim_{y_2\to 0-} f_2(y_2) = f_2(0)$, where $f_i(y)$ is the marginal density for the ith component $(i = 1,2)$. Our region R is $\{(\theta_1,\theta_2) \mid |\theta_1| \leq r_1, |\theta_2| \leq r_2\}$. As in example 6, $K_n(X(n)|\theta_1,\theta_2) = 0$ if $\theta_1 < \max(X_{11},\ldots,X_{1n})$, or $\theta_2 < \max(X_{21},\ldots,X_{2n})$; and $K_n(X(n)|\theta_1,\theta_2)$ is nonincreasing in each θ if $\theta_1 > \max(X_{11},\ldots,X_{1n})$ <u>and</u> $\theta_2 > \max(X_{21},\ldots,X_{2n})$. Also, $\lim_{n\to\infty} P_{\theta_1,\theta_2}[n\{\max(X_{11},\ldots,X_{1n})-\theta_1\} \leq y_1$ and

$$n\{\max(X_{21},\ldots,X_{2n})-\theta_2\} \leq y_2] = \lim_{n\to\infty} P_{\theta_1,\theta_2}[n\{\max(X_{11},\ldots,X_{1n})-\theta_1\}\leq y_1]$$

$$\times \lim_{n\to\infty} P_{\theta_1,\theta_2}[n\{\max(X_{21},\ldots,X_{2n})-\theta_2\} \leq y_2] =$$

$$= \begin{cases} 1 & \text{if } y_1, y_2 \text{ are both positive} \\ e^{y_1 f_1(0-)} & \text{if } y_1 < 0 \text{ and } y_2 > 0 \\ e^{y_2 f_2(0-)} & \text{if } y_1 > 0 \text{ and } y_2 < 0 \\ e^{y f_1(0-)+y_2 f_2(0-)} & \text{if } y_1 < 0 \text{ and } y_2 < 0. \end{cases}$$

It follows that $\left\{\max(X_{11},\ldots,X_{1n}) + \frac{r_1}{n}, \max(X_{21},\ldots,X_{2n}) + \frac{r_2}{n}\right\}$ is an m.p. estimator.

Example 8. $(X_{11},X_{21}), \ldots, (X_{1n},X_{2n})$ are n independent pairs, each with bivariate density $f(x_1,x_2|\theta) = \frac{1}{\pi\theta^2}$ if $x_1^2 + x_2^2 \leq \theta^2$, $f(x_1,x_2|\theta) = 0$ if $x_1^2 + x_2^2 > \theta^2$. Here $\theta > 0$, and $R(\theta) = (-r\theta, r\theta)$, $r > 0$. Here $K_n(X(n)|\theta) = 0$ if $\theta < \max\{\sqrt{X_{11}^2+X_{21}^2}, \ldots, \sqrt{X_{1n}^2+X_{2n}^2}\}$, $K_n(X(n)|\theta) = \{\frac{1}{\pi\theta^2}\}^n$ if $\theta \geq \max\{\sqrt{X_{11}^2+X_{21}^2}, \ldots, \sqrt{X_{1n}^2+X_{2n}^2}\}$. $\lim_{n\to\infty} P_\theta[n\{\max_i \sqrt{X_{1i}^2+X_{2i}^2} - \theta\} \leq -u] = e^{-2u/\theta}$ for $u > 0$. Thus $k(n) = n$, and the m.p. estimator is easily seen to be $\{1+\frac{r}{n}\} \max_i\{\sqrt{X_{1i}^2+X_{2i}^2}\}$.

References

[1] Anderson, T.W.: The integral of a symmetric unimodal function. Proc. Amer. Math. Soc., 6 (1955), 170-176.

[2] Anderson, T.W.: Introduction to multivariate analysis. Wiley, 1958, New York, N.Y.

[3] Cramér, H.: Mathematical methods of statistics. Princeton University Press, 1946, Princeton, N.J.

[4] Kaufman, S.: Asymptotic efficiency of the maximum likelihood estimator. Ann. Inst. Stat. Math., 18 (1966), 155-178.

[5] Kiefer, J. and Wolfowitz, J.: Consistency of the maximum likelihood estimator in the presence of infinitely many incidental parameters. Ann. Math. Stat., 27, No. 4 (1956), 887-906.

[6] Michel, R. and Pfanzagl, J.: The accuracy of the normal approximation for minimum contrast estimates. Zeitschr. f. Wahrscheinlichkeitsrechnung verw. Gebiete, 18 (1971), 73-84.

[7] Neyman, J.: Existence of consistent estimates of the directional parameter in a linear structural relation between two variables. Ann. Math. Stat., 22 (1951), 497-512.

[8] Roussas, G.G.: Contiguity of probability measures: some applications in statistics. Cambridge University Press, 1972.

[9] Wald, A.: Note on the consistency of the maximum likelihood estimate. Ann. Math. Stat., 20 (1949), 595-601.

[10] Weiss, L.: Asymptotic properties of maximum likelihood estimators in some nonstandard cases. Jour. Amer. Stat. Assoc., 66 (1971), 345-350.

[11] Weiss, L.: Asymptotic properties of maximum likelihood estimators in some nonstandard cases, II. Jour. Amer. Stat. Assoc., 68 (1973), 428-430.

[12] Weiss, L. and Wolfowitz, J.: Generalized maximum likelihood estimators. Teoriya Vyeroyatnostey, 11 (1966), 68-93.

[13] Weiss, L. and Wolfowitz, J.: Maximum probability estimators. Ann. Inst. Stat. Math., 19 (1967), 193-206.

[14] Weiss, L. and Wolfowitz, J.: Asymptotically minimax tests of composite hypotheses. Z. Wahrscheinlichkeitstheorie verw. Geb., 14 (1969), 161-168.

[15] Weiss, L. and Wolfowitz, J.: Maximum probability estimators and asymptotic sufficiency. Ann. Inst. Stat. Math., 22 (1970), 225-244.

[16] Weiss, L. and Wolfowitz, J.: Asymptotically efficient non-parametric estimators of location and scale parameters. Z. Wahrscheinlichkeitstheorie verw. Geb., 16 (1970), 134-150.

[17] Weiss, L. and Wolfowitz, J.: Maximum likelihood estimation of a translation parameter of a truncated distribution. Ann. Stat., 1 (1973), 944-947.

[18] Wolfowitz, J.: Estimation by the minimum distance method. Ann. Inst. Stat. Math., 5 (1953), 9-23.

[19] Wolfowitz, J.: The method of maximum likelihood and the Wald theory of decision functions. Indagationes Mathematicae, 15 (1953).

[20] Wolfowitz, J.: The minimum distance method. Ann. Math. Stat., 28 (1957), 75-88.

[21] Wolfowitz, J.: Asymptotic efficiency of the maximum likelihood estimator. Teoriya Vyeroyatnostey, 10 (1965), 267-281.

[22] Woodroofe, M.: Maximum likelihood estimation of a translation parameter of a truncated distribution. Ann. Math. Stat., 43 (1972), 113-122.

Vol. 310: B. Iversen, Generic Local Structure of the Morphisms in Commutative Algebra. IV, 108 pages. 1973. DM 16,-

Vol. 311: Conference on Commutative Algebra. Edited by J. W. Brewer and E. A. Rutter. VII, 251 pages. 1973. DM 22,-

Vol. 312: Symposium on Ordinary Differential Equations. Edited by W. A. Harris, Jr. and Y. Sibuya. VIII, 204 pages. 1973. DM 22,-

Vol. 313: K. Jörgens and J. Weidmann, Spectral Properties of Hamiltonian Operators. III, 140 pages. 1973. DM 16,-

Vol. 314: M. Deuring, Lectures on the Theory of Algebraic Functions of One Variable. VI, 151 pages. 1973. DM 16,-

Vol. 315: K. Bichteler, Integration Theory (with Special Attention to Vector Measures). VI, 357 pages. 1973. DM 26,-

Vol. 316: Symposium on Non-Well-Posed Problems and Logarithmic Convexity. Edited by R. J. Knops. V, 176 pages. 1973. DM 18,-

Vol. 317: Séminaire Bourbaki – vol. 1971/72. Exposés 400–417. IV, 361 pages. 1973. DM 26,-

Vol. 318: Recent Advances in Topological Dynamics. Edited by A. Beck, VIII, 285 pages. 1973. DM 24,-

Vol. 319: Conference on Group Theory. Edited by R. W. Gatterdam and K. W. Weston. V, 188 pages. 1973. DM 18,-

Vol. 320: Modular Functions of One Variable I. Edited by W. Kuyk. V, 195 pages. 1973. DM 18,-

Vol. 321: Séminaire de Probabilités VII. Edité par P. A. Meyer. VI, 322 pages. 1973. DM 26,-

Vol. 322: Nonlinear Problems in the Physical Sciences and Biology. Edited by I. Stakgold, D. D. Joseph and D. H. Sattinger. VIII, 357 pages. 1973. DM 26,-

Vol. 323: J. L. Lions, Perturbations Singulières dans les Problèmes aux Limites et en Contrôle Optimal. XII, 645 pages. 1973. DM 42,-

Vol. 324: K. Kreith, Oscillation Theory. VI, 109 pages. 1973. DM 16,-

Vol. 325: Ch.-Ch. Chou, La Transformation de Fourier Complexe et L'Equation de Convolution. IX, 137 pages. 1973. DM 16,-

Vol. 326: A. Robert, Elliptic Curves. VIII, 264 pages. 1973. DM 22,-

Vol. 327: E. Matlis, 1-Dimensional Cohen-Macaulay Rings. XII, 157 pages. 1973. DM 18,-

Vol. 328: J. R. Büchi and D. Siefkes, The Monadic Second Order Theory of All Countable Ordinals. VI, 217 pages. 1973. DM 20,-

Vol. 329: W. Trebels, Multipliers for (C, α)-Bounded Fourier Expansions in Banach Spaces and Approximation Theory. VII, 103 pages. 1973. DM 16,-

Vol. 330: Proceedings of the Second Japan-USSR Symposium on Probability Theory. Edited by G. Maruyama and Yu. V. Prokhorov. VI, 550 pages. 1973. DM 36,-

Vol. 331: Summer School on Topological Vector Spaces. Edited by L. Waelbroeck. VI, 226 pages. 1973. DM 20,-

Vol. 332: Séminaire Pierre Lelong (Analyse) Année 1971-1972. V, 131 pages. 1973. DM 16,-

Vol. 333: Numerische, insbesondere approximationstheoretische Behandlung von Funktionalgleichungen. Herausgegeben von R. Ansorge und W. Törnig. VI, 296 Seiten. 1973. DM 24,-

Vol. 334: F. Schweiger, The Metrical Theory of Jacobi-Perron Algorithm. V, 111 pages. 1973. DM 16,-

Vol. 335: H. Huck, R. Roitzsch, U. Simon, W. Vortisch, R. Walden, B. Wegner und W. Wendland, Beweismethoden der Differentialgeometrie im Großen. IX, 159 Seiten. 1973. DM 18,-

Vol. 336: L'Analyse Harmonique dans le Domaine Complexe. Edité par E. J. Akutowicz. VIII, 169 pages. 1973. DM 18,-

Vol. 337: Cambridge Summer School in Mathematical Logic. Edited by A. R. D. Mathias and H. Rogers. IX, 660 pages. 1973. DM 42,-

Vol: 338: J. Lindenstrauss and L. Tzafriri, Classical Banach Spaces. IX, 243 pages. 1973. DM 22,-

Vol. 339: G. Kempf, F. Knudsen, D. Mumford and B. Saint-Donat, Toroidal Embeddings I. VIII, 209 pages. 1973. DM 20,-

Vol. 340: Groupes de Monodromie en Géométrie Algébrique. (SGA 7 II). Par P. Deligne et N. Katz. X, 438 pages. 1973. DM 40,-

Vol. 341: Algebraic K-Theory I, Higher K-Theories. Edited by H. Bass. XV, 335 pages. 1973. DM 26,-

Vol. 342: Algebraic K-Theory II, "Classical" Algebraic K-Theory, and Connections with Arithmetic. Edited by H. Bass. XV, 527 pages. 1973. DM 36,-

Vol. 343: Algebraic K-Theory III, Hermitian K-Theory and Geometric Applications. Edited by H. Bass. XV, 572 pages. 1973. DM 38,-

Vol. 344: A. S. Troelstra (Editor), Metamathematical Investigation of Intuitionistic Arithmetic and Analysis. XVII, 485 pages. 1973. DM 34,-

Vol. 345: Proceedings of a Conference on Operator Theory. Edited by P. A. Fillmore. VI, 228 pages. 1973. DM 20,-

Vol. 346: Fučík et al., Spectral Analysis of Nonlinear Operators. II, 287 pages. 1973. DM 26,-

Vol. 347: J. M. Boardman and R. M. Vogt, Homotopy Invariant Algebraic Structures on Topological Spaces. X, 257 pages. 1973. DM 22,-

Vol. 348: A. M. Mathai and R. K. Saxena, Generalized Hypergeometric Functions with Applications in Statistics and Physical Sciences. VII, 314 pages. 1973. DM 26,-

Vol. 349: Modular Functions of One Variable II. Edited by W. Kuyk and P. Deligne. V, 598 pages. 1973. DM 38,-

Vol. 350: Modular Functions of One Variable III. Edited by W. Kuyk and J.-P. Serre. V, 350 pages. 1973. DM 26,-

Vol. 351: H. Tachikawa, Quasi-Frobenius Rings and Generalizations. XI, 172 pages. 1973. DM 18,-

Vol. 352: J. D. Fay, Theta Functions on Riemann Surfaces. V, 137 pages. 1973. DM 16,-

Vol. 353: Proceedings of the Conference on Orders, Group Rings and Related Topics. Organized by J. S. Hsia, M. L. Madan and T. G. Ralley. X, 224 pages. 1973. DM 20,-

Vol. 354: K. J. Devlin, Aspects of Constructibility. XII, 240 pages. 1973. DM 22,-

Vol. 355: M. Sion, A Theory of Semigroup Valued Measures. V, 140 pages. 1973. DM 16,-

Vol. 356: W. L. J. van der Kallen, Infinitesimally Central-Extensions of Chevalley Groups. VII, 147 pages. 1973. DM 16,-

Vol. 357: W. Borho, P. Gabriel und R. Rentschler, Primideale in Einhüllenden auflösbarer Lie-Algebren. V, 182 Seiten. 1973. DM 18,-

Vol. 358: F. L. Williams, Tensor Products of Principal Series Representations. VI, 132 pages. 1973. DM 16,-

Vol. 359: U. Stammbach, Homology in Group Theory. VIII, 183 pages. 1973. DM 18,-

Vol. 360: W. J. Padgett and R. L. Taylor, Laws of Large Numbers for Normed Linear Spaces and Certain Fréchet Spaces. VI, 111 pages. 1973. DM 16,-

Vol. 361: J. W. Schutz, Foundations of Special Relativity: Kinematic Axioms for Minkowski Space Time. XX, 314 pages. 1973. DM 26,-

Vol. 362: Proceedings of the Conference on Numerical Solution of Ordinary Differential Equations. Edited by D. Bettis. VIII, 490 pages. 1974. DM 34,-

Vol. 363: Conference on the Numerical Solution of Differential Equations. Edited by G. A. Watson. IX, 221 pages. 1974. DM 20,-

Vol. 364: Proceedings on Infinite Dimensional Holomorphy. Edited by T. L. Hayden and T. J. Suffridge. VII, 212 pages. 1974. DM 20,-

Vol. 365: R. P. Gilbert, Constructive Methods for Elliptic Equations. VII, 397 pages. 1974. DM 26,-

Vol. 366: R. Steinberg, Conjugacy Classes in Algebraic Groups (Notes by V. V. Deodhar). VI, 159 pages. 1974. DM 18,-

Vol. 367: K. Langmann und W. Lütkebohmert, Cousinverteilungen und Fortsetzungssätze. VI, 151 Seiten. 1974. DM 16,-

Vol. 368: R. J. Milgram, Unstable Homotopy from the Stable Point of View. V, 109 pages. 1974. DM 16,-

Vol. 369: Victoria Symposium on Nonstandard Analysis. Edited by A. Hurd and P. Loeb. XVIII, 339 pages. 1974. DM 26,-

Vol. 370: B. Mazur and W. Messing, Universal Extensions and One Dimensional Crystalline Cohomology. VII, 134 pages. 1974. DM 16,-